Volker Loeschcke (Ed.)

Genetic Constraints on Adaptive Evolution

With 52 Figures

Springer-Verlag
Berlin Heidelberg New York
London Paris Tokyo

Dr. VOLKER LOESCHCKE
Institute of Ecology and Genetics
University of Aarhus
Ny Munkegade
8000 Aarhus C
Denmark

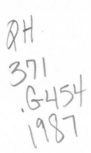

ISBN 3-540-17965-8 Springer-Verlag Berlin Heidelberg New York
ISBN 0-387-17965-8 Springer-Verlag New York Berlin Heidelberg

Library of Congress Cataloging in Publication Data. Genetic constraints on adaptive evolution. Bibliography: p. Includes index. Contents: A quantitative genetic perspective on adaptive evolution/J. S. F. Barker and R. H. Thomas – Genetic correlations/A. G. Clark – Genetic constraints on the evolution of phenotypic plasticity – [etc.] 1. Evolution. 2. Genetics. 3. Adaptation (Biology) I. Loeschcke, V. (Volker), 1950– . QH371.G454 1987 575.1 87-16518

Offsetprinting and bookbinding: Brühlsche Universitätsdruckerei, Giessen
2131/3130-543210

Preface

Fisher's fundamental theorem of natural selection contributed to the theoretical foundation of the Neo-Darwinian synthesis. Considered naively, however, it provided a rationale for neglecting genetic constraints on adaptive evolution, since the theorem concludes that evolution will increase the net fitness of a population. The amount of genetic variation in a population would, according to the theorem, only influence the rate of evolution, i.e., the time needed to reach the optimum. The fundamental theorem of natural selection, however, is based on simplifying assumptions that are often violated by a number of phenomena related to genetic constraints.

By genetic constraints on adaptive evolution is meant genetic factors which prevent the direct access to the optimum phenotype through natural selection. Such factors may include the genetic transmission, the genetic and the population structure, the mode of selection, the breeding system, and, last not least, the relation between genotype and phenotype. An important class of genetic constraints are due to linkage or to pleiotropy (genetic correlations). Other genetic constraints result from the absence of genetic variation, non-randomeness of spontaneous mutations, epistatic interactions, and the developmental organization resulting in non-linear mapping from genotype to phenotype.

This volume contains the papers presented at an Intecol symposium on "Evolutionary Constraints in Ecology", held at the 4th International Congress of Ecology at Syracuse, N.Y., on August 13, 1986. To complement these presentations, a few other contributions were taken up to broaden the scope. The symposium title was consequently changed to "Genetic Constraints on Adaptive Evolution" to stress the focus on genetic constraints on evolutionary processes.

The contributions stem from quantitative genetics, population genetics, ecological genetics and molecular genetics. Much attention is devoted to the organization and the maintenance of quantitative genetic variation, and to the intimate relation between genetics, ecology and evolution. The contributions to this volume also, in one way or another, attempt to open the black box that characterizes our poor knowledge on the processes connecting genotype and phenotype. The content of this box is evaluated in an evolutionary context with emphasis on how genetic aspects in the transmission of traits constrain adaptive evolution.

This volume is addressed to all evolutionary biologists and aims at making them suspicious towards a number of today's popular evolutionary concepts: the adaptionist program, the "advantage to the species" argument and optimization theory. Although useful in some contexts these ideas are based only on phenotypic characteristics and thus ignore genetic constraints on adaptive evolution.

Discussions and comments by Drs. Freddy Christiansen and Tom Fenchel during the preparation of this volume and editorial assistance by Ms. Marianne Szygenda are gratefully acknowledged.

Aarhus, July 1987   Volker Loeschcke

Contents

Chapter 8

Chapter 9

Contributors

You will find the addresses at the beginning of the respective contribution

Introduction: Genetic Constraints on Adaptive Evolution and the Evolution of Genetic Constraints

V. Loeschcke

Adaptive evolution can be defined simply as the process by which populations through natural selection increase their fitness in a given environment. Genetic constraints on adaptive evolution can then be understood as those genetic aspects that prevent or reduce the potential for natural selection to result in the most direct ascent of the mean phenotype to an optimum.

The simplest evolutionary constraint is the lack of genetic variation. Barker and Thomas (Chap. 1) argue that the lack of genetic variation is not an important evolutionary constraint, at least as long as a population is not exposed to rapid, major enviromental changes. They point to the ample evidence of genetic variation for fitness components and morphological, behavioral and physiological traits. To develop the quantitative genetics of these traits in natural populations, however, requires considerable ecological information. Barker and Thomas discuss the organization and maintenance of quantitative genetic variation, and critically review a number of models by carefully going through their assumptions. To explain the maintenance of quantitative genetic variation, they conclude, is not a problem of a deficiency of possible explanations, but rather a problem of their experimental evaluation and assessment of their relative importance.

The experimental evaluation, however, as Clark (Chap. 2) points out, is almost impossible as the hitherto available experimental techniques cannot be applied in natural populations without perturbations. When organisms are brought into the laboratory, this may result in gene by environment interactions which mask components of genetic variance for phenotypic characters of natural populations.

One way around the problem of inducing gene by environment interactions is to rear organisms under a variety of environmental conditions, in order to obtain the norm of reaction. This concept is taken up by Via (Chap. 3) in her contribution on the evolution of phenotypic plasticity. The rate of evolution of plasticity is determined by the genetic variation in reaction norms. She describes an experimental method for estimating the amount of genetic variation for phenotypic plasticity, where she views a character expressed in several environments as a set of genetically correlated character states. The correlation between character states is then employed as one of the parameters in a quantitative genetic model to determine the effects of different patterns of genetic variation in phenotypic plasticity on the evolution of plasticity.

From an ecological point of view most environmental variation is continuous. In quantitative genetics, however, the environment is generally treated as being discrete. Attempts to minimize environmental variation are made and what remains of variation in an experimental study is considered as "noise" (error variance). Van Noordwijk and Gebhardt (Chap. 4) sketch a framework for dealing with environmental parameters in quantitative genetics. A case study on reaction norms of developmental time and of weight at eclosion in *Drosophila mercatorum* is presented to show changes in the heritabilities of the traits and the genetic correlation structure due to genetic differences in the response to an environmental gradient.

Genetic correlations clearly make up an important class of evolutionary constraints which are themselves the product of evolution. Clark (Chap. 2) discusses to what extent models of mutation-selection balance can explain the observed levels of genetic correlations. Rose et al. (Chap. 5) compare mutation-selection balance models with optimization theory and the concept of antagonistic pleiotropy to discuss trade-offs in life history evolution. They quite generally question the value of optimization theory in life history evolution, and conclude that both antagonistic pleiotropy and mutation-selection balance may play a considerable role in the evolution of life history characters.

Christiansen (Chap. 6) choose another, novel approach to study the effects of pleiotropy in dynamical parameters of life history evolution. He considers the variation in a morphological character as the primary phenotypic variation and combines a simple description of the genetics of the character with a simple ecological description. Examples include the interplay between characters such as body size, reproductive effort and fecundity.

Scharloo (Chap. 7) exemplifies that genetic correlations are not absolute and inescapable: there are genes that act on several characters in the same direction, and genes that act on each character separately, as well as genes that act in an opposite direction. He emphasizes experiments showing that constructional features of the developmental system impose constraints on selection response and that such constraints can be built up by natural selection.

Developmental constraints as a bias in the production of variant phenotypes can also be expressed by non-randomness of mutations. Golding (Chap. 8) reviews and presents data that provide evidence for non-randomness of spontaneous mutations and compares the pattern of mutational changes with that observed within evolutionary substitutions.

Plant species have characteristics which to some extent distinguish their evolution from that of the animal species. For example, they have a more diverse breeding system which generally is assumed to be a result of adaptation to varying environments. For plants, the local environment is of crucial importance in adaptation, as they have to succeed in the spot where they live. Schaal and Leverich (Chap. 9) discuss experimental evidence which indicate genetic constraints on plant-adaptive evolution. They examine the interaction of genetic constraints with environmental variation. Much emphasis is devoted to the interaction of maternal effects and breeding system as well as the role of population structure in determining fitness and on the role of sexual selection in influencing adaptive evolution.

Chapter 1 A Quantitative Genetic Perspective on Adaptive Evolution

J. S. F. Barker and R. H. Thomas[1]

In order to understand adaptive evolution, we need to determine the effects of factors which shape and limit the expression of the essentially infinite possibilities of the genetic system. These factors are environment, and history as expressed in the present structure of the genetic system. Thus the study of adaptive evolution is the concern of both ecology and genetics, and we stress here the intimate relationship between them.

Ecological genetics has been a recognised branch of genetics for over 30 years, and evolutionary ecology is about the same age. The concepts of a genetic ecology are only now being developed, but even their need is being questioned (Stearns 1984). Clearly the hope for an integrated population biology (Sammeta and Levins 1970) has not yet been realised. Yet there have been significant developments—from the early theoretical studies of density-dependent and age-specific selection (Anderson and King 1970; King and Anderson 1971; Anderson 1971; Charlesworth 1971; Roughgarden 1971; Clarke 1972) to recent empirical studies of genetic variation, covariation and selection in natural populations (e.g. Price and Grant 1984; Grant 1985).

Until quite recently, ecological analysis was primarily in terms of individual phenotypes, at the level of the present and immediate past. While this approach has been very useful, we consider it as insufficient. Since present states are the outcome of evolution, the observations, processes and mechanisms of ecology are dependent on the evolving genotype, or as Dobzhansky (1973) more eloquently put it, "nothing in biology makes sense except in the light of evolution".

The study of adaptive evolution must include consideration of adaptation to the environment, differences among populations in both space and time, and fitness differences among individuals, all of which have a component that is genetic. While natural selection acts on phenotypes and our primary observations are at the level of the phenotype, the underlying phenomena are genetic. We cannot expect to understand the basis of phenotypic variation if we are ignorant of the genetics of the traits of interest.

With the focus on adaptation, it is important to clarify just what we mean. Krimbas (1984) gives a detailed review of the concept and the use of the term in evolutionary biology, while Wallace (1984) has replied to, and commented on, this review. The term is commonly used in three different ways, in the sense of (1) 'genetically adapted', (2) 'physiologically versatile or tolerant' or (3) 'developmentally flexible' (Maynard

---

[1] Department of Animal Science, University of New England, Armidale NSW 2351, Australia

Genetic Constraints on Adaptive Evolution
Ed. by V. Loeschcke
© Springer-Verlag Berlin Heidelberg 1987

Smith 1975). Here we are concerned with genetically adapted, although it should be noted that both physiological tolerance and developmental flexibility depend on the genotype, and in a sense are themselves genetic adaptations.

Again with the focus on adaptation, we take evolutionary constraints to be factors that prevent, or at least reduce the potential for, adaptive evolution (Michod 1984). In that paper, Michod also emphasizes the need to distinguish genotype and phenotype, using as descriptors for them the words 'intrinsic adaptiveness' and 'resulting fitness'. We do not see any value in this terminology and prefer the standard quantitative genetic terms of genotypic value and phenotypic value, applied in this case to fitness or to a trait that is a component of fitness. However, Michod does reemphasize the important point that differences among genotypes in their genotypic values are not necessarily directly related to their potential for evolutionary success, because the genotypic value may be masked either by the environment (in producing the phenotypic value) or by the process of transmission of genes from parent to offspring. Thus Michod defines any factor that leads to masking of genotypic value as a constraint on the evolutionary process, since the genotypes with highest genotypic values are not necessarily the ones that increase in frequency over evolutionary time.

Michod discusses one case (heterozygote superiority) where the masking is in the transmission process, and briefly mentions other examples, viz. frequency dependence, population density or growth rate and population age-structure, which are ecological variables that may act as constraints on evolution.

Other examples of constraints that are primarily genetic in operating through the transmission process include population genetic structure (Loveless and Hamrick 1984), coadapted gene complexes in fitness epistasis (Carson and Templeton 1984), genetic correlations among traits, expressed in quantitative genetic terms as the genetic covariance structure (Istock 1984) or as the specific genetic model of antagonistic pleiotropy (Rose 1982, 1983), the genetic architecture underlying the trait, particularly linkage relationships (Charlesworth and Charlesworth 1975; Templeton 1982) and the mating system (Templeton 1982).

In this chapter, we limit discussion to quantitative characters. Of necessity, knowledge of the quantitative genetics of natural populations will require considerable ecological information to allow meaningful interpretations of patterns of variation. Likewise, population structure, mediated in large part by ecological factors, will affect many aspects of genetic systems determining the expression of continuously varying characters, which in turn affect responses to selection.

We begin with some general comments on genetic variation for quantitative traits and then briefly discuss the relationship between qualitative and quantitative variation. Next we discuss the nature of genetic variation for quantitative traits and some of the properties and limitations of the methods used to describe that variation, particularly the concept of heritability. That done, we give an extended discussion of the organisation and maintenance of quantitative genetic variation, reviewing the theoretical models, such as mutation-selection balance, and antagonistic pleiotropy, before commenting on empirical work (and the lack thereof!). Finally, we close with a cautionary discussion of the limitations of conventional quantitative genetic analyses and a plea for pursuing the connections between genotypes and phenotypes.

Genetic Variation for Quantitative Traits

Early work in ecology, devoted to such matters as life history tables, population numbers, and rates of change in numbers, ignored genetic variation. Apart from Wright's (1931, 1932) shifting balance theory of evolution, which implicitly included interactions between genetic and ecological variables affecting the breeding structure of the population, recognition of these interactions has developed only over the last 35 years. We have noted the early theoretical work on density and age-structure effects, but much of the emphasis has been placed on the analysis of life history traits and development of a theory of life history evolution. In this, Istock (1984, and references therein) has been particularly prominent, in emphasising from his earliest review (1970) the significance of "The fundamental, and potentially selective, events... (of) births and deaths" and "the components of fitness (survivorship, longevity, time to age at first reproduction, and fertility) as the primary quantities which the theory must involve".

While components of fitness are the primary traits, attention recently has started to be given to morphological, behavioural and physiological traits which may have direct effects on fitness components, and workers in this area have rediscovered quantitative genetics. In addition, there has been a substantial development of quantitative genetic theory in relation to natural populations and evolution. Much of the credit for this must go to Lande (1976, 1980). We acknowledge this as a most valuable development, but in a later section, we will discuss come of the difficulties with present theory.

It should not be forgotten that there is already a vast body of theoretical and empirical work in quantitative genetics, much of it directed towards artificial selection and applications in plant and animal improvement. While there will be difficulties in extrapolating from experimental studies of laboratory organisms under controlled environmental conditions, or from results of artificial selection in domestic plants and animals, this body of knowledge should not be ignored and must provide a component of the scientific background to any consideration of quantitative traits in natural populations.

To some extent, there seems to be an element of surprise in the evolutionary biology literature that fitness components and other ecologically important traits may have a non-zero heritability, i.e. that there is genetic variation for these traits segregating in natural populations. The expectation that there should be little genetic variation for such traits presumably reflects an uncritical application of the Fundamental Theorem of Natural Selection (Fisher 1930). Fisher defined fitness in terms of the logarithmic rate of increase in number, calling it the 'Malthusian parameter of population increase' (m). Of course, ecologists will recognise this as the parameter (for a species) that is termed the intrinsic rate of increase, or the innate capacity for increase.

If population growth is not restricted in any way, population numbers will tend to increase (or decrease) exponentially, so that

$$\frac{1}{N}\frac{dN}{dt} = \frac{d\log_e N}{dt} = \text{constant } (m).$$

Hence, after t generations,

$$N_t = N_0 e^{mt},$$

where N_t is population number at generation t and N_0 the initial population number.

Given this definition of fitness, Fisher showed that for certain idealised conditions, the rate of change in the mean fitness of a population is given as:

$$\frac{d\overline{m}}{dt} = V_A(m),$$

where $V_A(m)$ is the additive genetic variance in fitness. Since a variance must be positive, $\frac{d\overline{m}}{dt}$ must be positive, so that fitness defined in these terms will tend always to increase.

In a constant environment with no mutation, natural selection would exhaust the additive genetic variance for fitness, and there could then be no further evolutionary change. Fisher recognised that the environment, both physical and biotic, does not remain constant. Further, many of his idealised conditions are not appropriate to the real world, e.g. that the age distribution would remain constant (Charlesworth 1970), while other assumptions were not explicit, e.g. constant genotype fitnesses and weak selection (Crow and Kimura 1970; Charlesworth 1980) so lack of genetic variation for fitness is not inevitable.

In fact, there is ample evidence of genetic variation for fitness components and morphological, behavioural and physiological traits, either as direct estimates of heritability or from results of artificial selection experiments, and this evidence is not only to be found in the plant and animal breeding literature. Lewontin (1974a) noted that "it is a commonplace of Drosophila population genetics that 'anything can be selected for' in a non-inbred population", and listed a number of examples "to provide an impression of the extraordinary variety of physiological, behavioral and morphogenetic processes for which there is genetic variance in substantial quantities in outbreeding Drosophila populations". Cade (1984) has reviewed evidence for genetic variation in traits associated with sexual behaviour and reproduction.

Thus the simplest possible evolutionary constraint, viz. lack of genetic variation, would appear not to be important, although it is possible that the genetic variation is itself constrained through pleiotropy or genetic correlations and not available for further evolutionary change. In addition, the genetic variation that would allow a population to survive following a rapid, major environmental change may well be lacking (Bradshaw 1984).

Qualitative and Quantitative Genetic Variation

Experimental population genetics for nearly 20 years has attempted to analyse genetic variation and to interpret the basis of the large amounts of genetic polymorphism for structural enzyme loci in natural populations. Resolution of the continuing debate on the relative importance of selective maintenance of polymorphism and of neutral variation will depend, at least partly, on establishing the adaptive significance of particular enzyme polymorphisms (Clarke 1975; Koehn 1978), and substantial progress has been made in the analysis of some enzyme loci, e.g. *alcohol dehydrogenase* in *Drosophila melanogaster* (Van Delden 1982; Clarke and Whitehead 1984), *leucine aminopeptidase* in *Mytilus edulis* (Hilbish and Koehn 1985) and *phosphoglucose isomerase* in *Colias* butterflies (Watt et al. 1985).

It is equally important that we understand the nature of genetic variation for quantitative traits, the forces that may contribute to maintaining this variation, and the reasons for differences among populations in the variation for any particular trait. This cannot yet be done in terms of individual loci, and, as argued by Lewontin (1984), it cannot be assumed that there is any necessary connection between population differentiation for quantitative traits and for gene frequencies at qualitative loci. That is, we should not draw conclusions about possible variation among populations in quantitative traits from studies of qualitative variation. (This conclusion holds, notwithstanding the discussion about inferences that might be drawn from comparisons of between-population variation for quantitative traits and gene frequencies (Rogers 1986; Felsenstein 1986; Lewontin 1986).

However, a number of empirical studies have found an inverse relationship between genic heterozygosity (estimated from the electrophoretic assay of protein-coding loci) and the phenotypic variance of various morphological and other traits (see reviews in Mitton and Grant 1984; Livshits and Kobyliansky 1985; Zink et al. 1985). These findings are usually explained in terms of Lerner's (1954) concept of developmental homeostasis, that phenotypic variability is negatively correlated with the level of heterozygosity, and that heterozygotes are favoured by selection. Chakraborty and Ryman (1983) argue that the negative relationship between genic heterozygosity and phenotypic variance can be explained by additivity of gene effects that control the quantitative character, and that there is no need to invoke any selection hypothesis. However, their analysis makes two important (but not explicitly stated) assumptions, so that the explanation may not be so simple. Firstly, they assume that electrophoretic loci are a random sample of the genome, or at least that part of the genome affecting polygenic variation. Secondly, a constant environmental component of variance is assumed so that the decrease in phenotypic variance with an increase in heterozygosity is suggested to be a function solely of the decrease in genotypic variance. However, the concept of developmental homeostasis includes the notion that heterozygotes are better buffered against environmental deviations, so that the environmental component of variance also would be expected to decrease as heterozygosity increased. Clearly analysis at the level of phenotypic variance is not sufficient to resolve the issue, and a direct study of the relationship between genic heterozygosity and additive and non-additive genetic variances would be more informative. We know of only one such study (Gunawan 1981), where for nine isofemale lines derived from one population of *Drosophila buzzatii*, the regression coefficient of additive genetic variance for abdominal bristle number on heterozygosity at enzyme loci was positive, but not significantly different from zero. Further and more detailed studies are needed.

Nature of Genetic Variation for Quantitative Traits

The phenotypes of individuals may be measured and the phenotypic variance (V_P) estimated. As the differences among individuals in their phenotypes are due to differences in their genotypes and in their environments, the phenotypic variance of a population is due to genetic variance (V_G), to environmental variance (V_E) and to genotype x environment interaction variance. Thus the phenotypic variance for some trait gives no necessary guide to the amount of genetic variation, let alone the nature of that variation, and theories and concepts based only on phenotypic variation may be quite misleading. Phenotypic variation is necessary for evolutionary change, but it is not sufficient.

In theory, the genetic variance may be partitioned further (Falconer 1981) into additive (V_A), dominance (V_D) and epistatic or interaction (V_I) variance, the latter two described together as non-additive. Knowledge of the relative magnitudes of these three genetic components would give some understanding of the nature of genetic variation, but it must be recognised that this is a statistical description of variation and not one based on direct analysis of gene action. The ratio of the additive genetic variance to the phenotypic variance is termed the heritability of the trait, and has two functions: one descriptive, to describe the amount of additive genetic variation, and one predictive, to predict the expected response to directional selection.

However, there are very real constraints on the application of this quantitative genetic theory to natural populations. Firstly, estimation of heritability depends on measuring correlations among relatives. Thus only those species can be studied where individuals can be marked and identified, and breeding histories followed to identify relatives. Plants will offer some advantages (Lawrence 1984), with the possibility of controlled pollination and experimental randomization, but there are still many sources of potential bias (Mitchell-Olds and Rutledge 1986). Secondly, even where heritabilities can be estimated in the natural population, any further attempt to partition the variation in order to gain some knowledge of the non-additive variation is not likely to be very informative. Even in laboratory populations under controlled environmental conditions, and using large, well-designed experiments with many different types of relatives, attempts to describe non-additive quantitative genetic variation have not been very successful (Barker 1974). In addition, good estimates of the genetic variance components still would only be statistical estimates; they would not define the genetics of the trait in terms of specific effects of individual loci.

There is the possibility of misunderstanding because of a failure to recognise what heritability is not. It is not some sort of fundamental constant, i.e. a fixed value for a given trait. It is a ratio that depends on both additive genetic and environmental variance. Thus the heritability of a trait may differ between populations either because of differences in the amount of additive genetic variation, or of environmental variation, or both. This is of course particularly pertinent where laboratory studies are extrapolated to the natural population from which individuals were sampled. With laboratory control of environmental variation, and hence reduction in V_E as compared with that in the natural population, estimated heritabilities may be expected to be higher than those applicable in the natural population (Prout 1958; Atkinson 1979). In fact, the problem of extrapolation may be even more acute, as found by Prout

(1958) for wing length of *D. melanogaster*. When both wild-caught and laboratory reared males were used as sires in full- and half-sib mating designs, substantial additive genetic variance was found, yet for wild-caught males as sires, the sire-offspring covariance was negative (i.e. zero heritability). However, it may often be the case that it is not the precise value of the heritability that is of interest, but rather the magnitude of additive genetic variance and particularly whether it is non-zero. Assuming a large sample initially taken from the natural population and an appropriate experimental design to minimise the standard error of heritability, a non-zero additive genetic variance estimate in the laboratory at least indicates the existence of such variance in the natural population. It must still be recognised that the magnitude of this variance in the laboratory may be different from that in the natural population because the average effects of genes may differ in the two environments or because of genotype x environment interaction.

Organisation and Maintenance of Quantitative Genetic Variation

Biologists have long recognised the existence of genetic connections between characters; when one is selected on, the other also responds in a predictable manner. These are the 'mysterious laws of correlation' which were so necessary to Darwin's theory (1859), and so perplexing to him, lacking as he did an adequate theory of inheritance. It could be argued that to date we have made little progress in understanding the nature of genetic covariances of groups of characters. This problem is one deserving more attention, and Clark (this Vol.) considers it in terms of attempts to explain senescence. Considerably more progress has been made in understanding the effects of selection on phenotypic evolution, given a genetic covariance structure. Here Lande's analyses (Lande 1979; Lande and Arnold 1983) are central and must be considered separately from his genetic analyses (which will be covered later in this section). The fundamental difficulty with these sorts of phenotypic models is that genetic parameters are time-invariant constants. Hence we get no insight into the evolution of the covariance structure itself. Over the past 10 years a veritable industry has sprung up to measure genetic and phenotypic covariance structures for a variety of classes of characters in a wide array of taxa, both in nature and in the laboratory. We make no attempt to review this literature here. Studies of geographic variation in correlation structure have been especially interesting, and two examples of this work are those of Dingle and his colleagues on *Oncopeltis* (Baldwin and Dingle 1986; Dingle 1981; Dingle and Baldwin 1983) and Via on *Liriomyza* (Via 1984a,b). Via's work is the first attempt in natural populations to consider the correlation of expression of a character in two environments as a means of analysing genotype-environment interactions (Falconer 1952; Robertson 1959). This approach makes it possible to analyse genotype-environment interaction in terms of multivariate models of phenotypic evolution (Via and Lande 1985), an important advance in the study of evolution in heterogeneous environments.

Now we consider the maintenance of quantitative genetic variation. First we make some general comments and then look at specific models for explaining the maintenance of variation and the controversy surrounding them.

The controversy in population genetics over the relative importance of selective maintenance of polymorphism and of neutral variation has had little impact in relation to quantitative variation and it is generally accepted that the neutral theory of molecular evolution does not apply to morphological evolution (Nei and Graur 1984). For quantitative variation, it is assumed that the variation is maintained by some form of balancing selection or balance between selection and mutation or migration (reviewed by Grant and Price 1981 in an ecological context).

Many morphological and behavioural traits of ecological interest, and fitness components, are assumed to have intermediate optima, i.e. individuals with values near the population mean have highest fitness, and fitness decreases as the values deviate in either direction from the mean. Selection thus favours intermediates and this is referred to as stabilising selection. If stabilising selection is so pervasive, it will be important to consider its effects on genetic variation. However, as Falconer (1981) has emphasised, a careful distinction needs to be made as to whether the apparent stabilising selection is real or spurious. If it is real, the trait is itself selected, and the value of the trait is a direct cause of fitness. It it is spurious, intermediates still have the highest fitness, but the connection with fitness is through pleiotropic effects of genes.

The effects of apparent stabilising selection on genetic variation then depend on whether it is real or spurious. If spurious, the effect cannot be determined, although Robertson (1956) has suggested one model where genetic variance would be maintained. Real stabilising selection favours those genotypes that show the least variability and it reduces genetic variation (Falconer 1981). There is an obvious need for appropriate studies in natural populations to determine the extent to which stabilising selection is real or spurious, and to determine if generalisations are possible with regard to classes of traits. Observed high levels of genetic variation do not mean that real stabilising selection is not important, as additive genetic variance can be maintained by mutation even with strong stabilising selection (Lande 1976). Until quite recently (Franklin 1982; Hill 1982), quantitative geneticists have ignored the possible importance of mutation in contributing to genetic variance of quantitative traits and responses to directional selection. Further, the optimum cannot be expected to be stable over evolutionary time, and will shift with changes in the environment. This imposed directional selection, sometimes for an increase in the optimum, sometimes for a decrease, could contribute to the maintenance of genetic variation. Whether it would do so or not depends on the strength and time scale of particular episodes of directional selection.

Milkman (1982) has pointed out that all earlier models of stabilising selection which lead to decreases in genetic variation assumed infinite population size. In finite populations, where fitness itself is the trait of interest, and assuming a very large number of loci affecting it, he argued that alleles can be considered as selectively neutral, i.e. selective forces are too weak relative to genetic drift to drive the population towards homozygosity.

Mutation-Selection Models and Phenotypic Models: Assumptions and Tests

As mentioned earlier, Lande (1976, 1980) is largely responsible for the current re-naissance in theoretical quantitative genetics as applied to natural populations. His models attempt to show how mutation and stabilising selection can maintain obser-ved levels of equilibrium genetic variation in quantitative traits and how weak direc-tional selection affects their long-term evolution. Lande achieves his results by ignor-ing as irrelevant the genetic details of the construction of phenotypes, while retaining consideration of the dynamics of genotypes. This makes his mathematics both tract-able and general. The dynamics of selection can be followed in the relatively low dimensional space of phenotypes rather than that of genotypes (Lewontin 1985). There are, of course, a number of assumptions involved in permitting these simplifi-cations. The following comments point out these assumptions and the consequences of their acceptance or rejection, and show how models with quite different genetic bases can yield similar predictions. Much of the discussion will be relevant to pheno-typic models of selection in general, as they often make similar assumptions (Lloyd 1977).

Clegg and Epperson (1985) have listed what they see as the assumptions of a class of models of stabilising selection-mutation which Turelli (1984) refers to as K-L-F models after Kimura (1965) (who really started it all), Lande (1976) and Fleming (1979). There are disagreements between Lande (pers. comm.) and Turelli (pers. comm.) regarding who is assuming what. Turelli agrees with Clegg and Epperson's (1985) statement of the assumptions, while Lande disagrees with several in the list. Most of the controversy derives from four critical areas for which empirical evidence is inadequate. These areas, as reviewed in Turelli (1984), are: per locus mutation rates, the genetic changes underlying observed mutation, effects attributable to indi-vidual loci and the nature and intensity of selection. We will use Clegg and Epperson's (1985) list as a basis for discussion. These assumptions, in addition to the usual many genes, each of small effect, are:

"1. The distribution of phenotypes is normal.
2. Population size is effectively infinite.
3. There is a continuum of alleles at each locus.
4. Each allele at each (interchangeable) locus contributes an additive genetic effect (within and between loci) plus a normally distributed, genotype-independent, ad-ditive environmental effect.
5. The change in the effect of a gene by mutation has a normal distribution with mean zero.
6. The selective value of phenotypes has either a Gaussian or 'quadratic' (in the Ki-mura model) distribution.
7. Linkage between loci is 'loose'."

We will consider each of these assumptions in turn.

A normal distribution of phenotypes is a very common empirical observation. If a trait is non-normal in the initial scale of measurement, it can often be made normal by means of a suitable transformation. Thus this assumption is usually justified when dealing with traits showing continuous (or nearly continuous) variation. Even in the case of discrete phenotypes produced by a 'developmental switch' mechanism (e.g. Lively 1986) the assumption of normal distribution may still be appropriate if we refer it to the threshold of the developmental switch itself. The variation in the threshold (sensitivity) may be continuous and normally distributed. Normality of the phenotype distribution is quite robust even to violation ot the assumption of many genes, each of small effect (e.g. Hammond and James 1970). However, with major genes segregating, there may be significant violation of normality if directional dominance is present.

The next assumption, of an effectively infinite population size, is in our view considerably more problematic. As ecologists are well aware, the estimation of population size is a difficult task. Estimation of the genetically effective population size is even more difficult, and fraught with many uncertainties. There are few estimates, good or otherwise, of effective population size, but a large proportion of them suggest population sizes small enough for genetic drift to be a significant evolutionary force (Thomas, unpubl.). Many species, at some times since their origin, are likely to have had small effective population sizes. Depending on the circumstances of these reductions in population size, a genetic architecture of polygenic traits could be produced that would violate the assumption of many genes, each of small effect (Templeton 1981). Given the paucity of available evidence, this area remains a contentious one (Lande 1981). More investigations of population structure and effective population size are badly needed, and collaboration between ecologists and geneticists would facilitate progress.

The third assumption, of a continuum of alleles at each locus, follows the analyses of Crow and Kimura (1964) and is motivated by the idea that there is possible at each locus a large number of alleles due to mutation. Combining this assumption with that of an effectively infinite population size implies that a large number of alleles are segregating in a population at a given time. Kimura's (1965) analysis considered the evolution of the distribution of allelic effects at a single locus in linkage equilibrium. He extrapolated this analysis to an n locus situation and obtained the result of an approximately Gaussian distribution of allelic effects at each locus contributing to a character (assumption 4). Turelli (1984, 1986) has critically examined Kimura's result and Lande's (1976, 1980) incorporation of it as an assumption into his models. In Lande's (1980) generalisation of these models to include pleiotropy, there is the explicit assumption that in general each locus, for every character it affects, can produce an unbounded continuum of effects. Turelli (1986) says, "this level of genetic flexibility is a strong assumption". Unfortunately it is a consequence of the assumption necessary to make the model mathematically tractable. Turelli (1985) has shown that assuming less genetic flexibility leads to large effects on quantitative prediction. Depending on the nature of the pleiotropic selection, the equilibrium genetic variance for a single trait can be greatly over- or underestimated.

The fifth assumption, that the distribution of mutation effects at a locus is normal, is based on biological arguments that are open to question. The multivariate density function describing the multilocus gametic composition of the population is approximately Gaussian only if the variance of mutation effects is small relative to the variance of effects for currently segregating alleles (Lande 1976; Turelli 1984, 1985).

Turelli (1984) has considered the consequences of this assumption in great detail and concluded that it implies per locus mutation rates of greater than 10^{-4}. Mutation rates of this magnitude do not conform to traditional estimates of $<10^{-5}$. However, in fairness it must be said that it is not at all clear that these traditional estimates of per locus mutation rates apply to typical loci affecting quantitative traits. To account for the observed total mutation rate of approximately 10^{-2} per gamete per generation per character, either one has to accept higher per locus mutation rates, or more loci than there is direct evidence for, or some 'unusual', but not necessarily uncommon, mutation process, or any or all of these. Convincing evidence is lacking for all these areas.

Stabilising selection is incorporated into the models by assuming a Gaussian or quadratic fitness function. This assumption stems from the observation that many phenotypic characters seem to have optimal values.

Lande's (1976) analyses of his model show that linkage disequilibrium introduces only small changes in the approximation for the equilibrium genetic variance. However, Turelli (1984, 1986) notes that Lande's (1976) multilocus predicted equilibrium genetic variance is only trivially different from that produced by considering each locus separately.

Turelli (1984, 1985, 1986) has taken two paths in trying to make mutation-selection balance models that he considers more biologically realistic than the K-L-F models. Firstly, he proposed an alternative approximation for the distribution of effects of new mutants in the continuum-of-alleles model. His 'house-of-cards' approximation assumes a new mutant has a phenotypic effect independent of the average effect of alleles segregating at the locus. This follows from the acceptance of traditional per locus mutation rate estimates, which in turn results in the expectation that the genetic variance of mutation-induced effects is large relative to that caused by currently segregating alleles. As Turelli (1986) notes, "this is analogous to the classical strong selection, low mutation rate approximation for mutation-selection balance in which essentially all deleterious mutants arise from 'wild type' alleles". He finds this approximation is usually accurate for low mutation rates per locus when selection is not extremely weak. Secondly, Turelli considered the effect of a finite number of alleles at each of the loci influencing a trait. Using low mutation rates per locus and the house-of-cards approximation, he obtained similar results for the continuum-of-alleles model and a model with three alleles, implying that the relative magnitudes of mutation effects and of equilibrium genetic variance is the crucial factor, rather than the number of alleles per locus. However, Barton (1986) has shown that the equilibrium genetic variance is critically dependent on the history of the population, and that it cannot be predicted from the observed distribution of the character. He therefore concluded that only a crude description of the evolution of continuous characters will be possible without detailed knowledge of their genetic basis.

All of these models predict the maintenance of considerable heritable genetic variation for polygenic traits by mutation-selection balance, but except for that of Lande (1980), they assume that selection acts on the relevant loci only through the single character being studied. More realistically we might expect that these loci will have unobserved pleiotropic effects on other characters under selection. Turelli (1985) considered the consequences of these hidden pleiotropic effects. The results are complex but the principle conclusion is that, given hidden pleiotropic effects, accurate predictions concerning mutation-selection balance for polygenic traits will be very difficult or impossible to achieve.

It seems obvious that any genetic explanation of the maintenance of genetic variation in polygenic traits must include mutation and selection. However, a number of models making plausible genetic assumptions, but not including mutation-stabilising selection balance, also predict the maintenance of variation. Gillespie (1984) examined a situation where alleles contribute to the fitness of the organism in two ways. Firstly, alleles are assumed to contribute additively to the phenotype just as in the mutation-selection balance models. The fitness of an individual's phenotype follows from the assumptions of Wright's (1969) optimum model in which there are multiple fitness peaks resulting from pleiotropy. Secondly, an individual homozygous at a locus is assumed to have lower fitness than another individual with the same phenotype but heterozygous at that locus, i.e. there is pleiotropic overdominance. Genetic variation is maintained under these conditions.

Other possible explanations for the maintenance of quantitative genetic variation include frequency-dependent selection, density-dependent selection (if it is also frequency-dependent), spatial and temporal environmental heterogeneity (another form of frequency dependence), genotype-environment interactions, and specifically for traits that are components of fitness, antagonistic pleiotropy (which implies overdominance for fitness itself). The problem is not one of a deficiency of *possible* explanations, but rather of their experimental evaluation and assessment of their relative importance in natural populations. Rose, Service and Hutchinson (this Vol.) discuss some of these explanations in the context of the evolution of life histories and the need for empirical testing, and report on experiments which suggest that several mechanisms are acting to varying extents on different characters. While it might be argued that such empirical testing is difficult, if not impossible, in the natural populations themselves, carefully designed laboratory experiments can tell us what is feasible. However, there have been few such experiments, their results are equivocal with respect to theory, and experiments lag far behind the theoretical developments.

The relationship between genetic and environmental variability was studied by Mackay (1981), using three quantitative traits of *Drosophila melanogaster*. For two traits (body weight and sternopleural bristle number), additive genetic variance increased over time to become significantly greater in the variable environments than in a control population, but for the third trait (abdominal bristle number), there were no clear differences. As genotype-environment interactions were not significant, it was concluded that adaptation to a variable environment is by selection for heterozygosity, rather than by diversifying selection for different alleles in alternate niches. In a second study with *D. melanogaster*, Tachida and Mukai (1985) used strains homozygous or heterozygous for different second chromosomes extracted from a wild population, and measured viability in nine different environments. The variance component for genotype-environment interaction was significantly different from zero, and that for homozygotes was significantly greater than for heterozygotes. From a further partitioning of the interaction variance, the authors concluded that diversifying selection was acting to maintain additive genetic variation for viability. Dawson and Riddle (1983) undertook a long-term natural selection experiment with

Tribolium confusum and *T. castaneum* to test whether genetic variability in fitness was maintained by environmental diversity. Although they found significant genotype x environment interactions for a variety of growth traits and fitness components, the results did not support the hypothesis that genetic variation is maintained by diversifying selection in a heterogeneous environment. They suggested that pleiotropic effects of genes on different fitness components (expressed as negative genetic correlations among the components) was the most likely explanation of their results.

This concept of antagonistic pleiotropy maintaining polymorphism is not new (Caspari 1950), but it has recently received considerable theoretical (Lande 1982; Rose 1982, 1983, 1985) and experimental attention (Rose and Charlesworth 1980, 1981; Rose 1984; Luckinbill et al. 1984; Luckinbill and Clare 1985), particularly in the context of life history evolution. For example, Rose (1982, 1985) considered pleiotropically connected traits on which selection is acting in different directions. For both discrete and overlapping generations, he found that protected polymorphisms could be generated without overdominance in gene effects on individual life history traits. Again the empirical results in the above papers, and in other studies quoted by them, are contradictory, but some possible reasons for the differences have been identified in terms of particular genotype x environment interactions that were important only in some of the experiments (Clare and Luckinbill 1985; Service and Rose 1985).

As far as the maintenance of genetic variation is concerned, there are contradictions within the empirical results, and discrepancies between these results and theoretical expectations. Further work, particularly experimental, will be needed before a clearer picture can emerge.

Explaining observed levels of heritable quantitative variation will not be easy, and there should be no expectation that only one mechanism will be adequate. Considering different traits within species and different species, there are bound to be different mechanisms. Distinguishing the contributions of different mechanisms will be difficult, given the subtlety of the differences in the predictions made by the various models, and given the difficulties in estimating the necessary genetic parameters.

However, there is another aspect to genetic variation that may be very important for evolutionary change, and it seems not to have been considered in this context. This is the genetic variation that is not expressed because of the canalisation of development, a reduction in phenotypic variance between two developmental thresholds (Waddington 1957). Many developmental processes apparently are canalised, as individuals in natural populations tend to have a uniform morphology, particularly in their major features. On the other hand, individuals homozygous for a recessive mutant or carrying a dominant mutant that affects some major morphological feature are often phenotypically very variable. The reason for this difference, according to Waddington (1957), is that the normal wild-type morphology results from an evolved canalisation due to stabilising selection, while the mutant morphology has not been subjected to such selection. Selection against variability in the presence of a major mutant can be successful, the change can be quite rapid and it can lead to a new phenotypic norm (Rendel 1967). Thus when canalisation is de-regulated by such a mutant or by environmental stress (Waddington 1953), genetic variability is exposed, and quite major developmental changes may occur. In the rapidly changing environments to which some populations are being exposed as a result of human activities, this possibility should be considered in relation to population differentiation. It also might be noted that such de-regulation of canalisation may provide a basis for speciation.

As noted previously, the traits that have been considered in the quantitative genetic analysis of natural populations have mainly been fitness components and morphological traits affecting fitness, i.e. the phenotypic end-points of developmental processes chosen for study primarily because they could be readily defined and objectively measured. But the issue of canalisation should remind us that the constraints of concern may be developmental (Maynard Smith et al. 1985).

Quantitative Genetic Analyses

It will be apparent from the preceding discussion that conventional quantitative genetic analysis does not mean an analysis in terms of genes and gene effects, but rather the statistical analysis of phenotypes in terms of means, variances and covariances. As such, it is based on a set of assumptions about the genetic and environmental determination of the quantitative phenotype. Some of these assumptions have been stated explicitly in the development of the theory, some are believed to be true and some are known not to be true (at least sometimes). The effects of relaxing some of the assumptions have been investigated, but we do not intend here to review these effects. What we do wish to emphasise is that before undertaking quantitative analyses, and particularly before interpreting an analysis, the assumptions should be recognised, and their possible effects considered. As this is a cautionary tale, we list the assumptions as follows (noting that not all quantitative analyses make all these assumptions):

1. Genes affecting the trait are inherited as normal Mendelian genes.
2. Quantitative traits are determined by the joint effects of the environment and a large number of unlinked, non-interacting loci with effects on the phenotype that are equal, additive and small (in relation to the effects of the environment).
3. The population is random mating, with genotype frequencies in Hardy-Weinberg equilibrium, i.e. there is no inbreeding, assortative mating, selection or migration.
4. There is no genotype-environment correlation.
5. There is no genotype x environment interaction.

Although not explicitly stated as part of quantitative genetic models, other implicit assumptions commonly include:

1. There are no effects of competition or cooperation among individuals within the population, or between these individuals and members of other species.
2. The genetic variation is due to structural genes, i.e. genes coding for proteins.

The latter assumptions may not matter in theories or models where alternative alleles at any locus are assigned either a '+' or a '-' effect, or even a continuum of allelic effects, but it may restrict our thinking about quantitative variation. Further, the second has been found not to be true in one case (Frankham 1980) where responses to selection for a quantitative character (abdominal bristle number in *Drosophila melanogaster*) were due to changes in copy number of the ribosomal RNA tandon (a multiple copy, tandemly repeated structure).

The important genetic variation for quantitative traits may be that at regulatory loci (Barker 1985), and Wilson (1976) has suggested that changes in gene regulation may be of critical importance in evolution, as possibly the basis of major adaptive changes.

Finally, although not assumptions, it should be noted that:

1. Any analysis is specific to a particular population in a particular environment at a particular time.
2. The heritability of a trait, the usual primary estimate, refers only to variation among individuals within the population studied; it tells us nothing about variation among populations.

The quantitative genetic models that have been developed in recent years have been discussed above. These models depend on essentially the same assumptions, and the major difference as compared with earlier theory is in the emphasis on the type of selection considered. Early theory related to a description of variation and the use of heritability to predict expected responses to directional selection, as in animal and plant breeding selection programs. The more recent theory has its emphasis on natural populations and stabilising selection.

Nevertheless, whether we are attempting to increase the efficiency of artificial selection, or to understand past natural selection effects on the magnitude and maintenance of genetic variation in natural populations, our ultimate requirement would be a full description of the genetic variation, i.e. the number of loci affecting the trait, the number of alleles segregating at each locus, their frequencies, the nature and magnitudes of gene effects, the degree of dominance, the nature and magnitude of interlocus interactions, linkage relations, mutation rates and the pleiotropic effects on other traits of loci affecting the trait of interest.

We do not now and we may not ever have such complete information. Thus experimental quantitative genetics may well remain the applied mathematics and statistics of variation and estimation, and would be better described as "quantitative phenetics".

The quantitative genetic models tell us what is possible, and their further exploration should help to direct our search for understanding. But that search must depend also on analysis at the level of the gene, of the DNA itself, on gene effects and on the biochemical, physiological and developmental pathways from the genes to the phenotypes.

Lewontin (1974b) has emphasised that the analysis of variance and the partitioning of phenotypic variance into genetic and environmental components does not produce an analysis of causes, nor does it specify functional relations. As he points out, "The real object of study . . . [is] the *norm of reaction,* which is a table of correspondence between phenotype, on the one hand, and genotype-environment combinations on the other". The relative phenotypic values of different genotypes depend on the particular environment to which each has been exposed. It is not sufficient to say that all the genotypes (individuals) are of the one population and therefore have been exposed to the same environment. No two individuals can have been exposed to *exactly* the same set and sequence of environmental factors. Equally, individuals that do not differ phenotypically, or populations that do not differ in mean phenotype for a particular trait, may still be different genetically. For example, Yoo (1980a) selected for abdominal bristle number in six replicate lines of *D. melanogaster*, taken from the same population and selected using large population size under the same experimental and environmental conditions. After 85 generations of selection, the lines were in two sets, each of three lines, with a similar mean phenotype within each set. But within each of these phenotypically similar sets, the lines were very different genetically (Yoo 1980b).

In order to analyse quantitative genetic variation, and to increase our understanding of the nature of this variation, it is essential to consider simultaneously both genetic and environmental effects, and to focus on the norm of reaction. Further, as has been emphasised in the context of quantitative traits in animal breeding (Barker 1985), our observations are at the level of the phenotype, which is partly dependent on an unknown genotype, and the pathways between the genotype and the phenotype are largely inside a closed and unknown black box. What is needed is a two-directional approach to opening that box — one from research to understand gene structure, control and regulation, and gene action and interaction, the other working back from the phenotype into the physiology, biochemistry and development of the organism.

Clegg and Epperson (1985) noted "It is apparent that the study of quantitative characters has experienced a renaissance in recent years". However, their review essentially considered only the renaissance of work on the theory. The renaissance in empirical studies is yet to come, but appears to have begun, and it will certainly be fostered by a greater appreciation of the need to unravel the pathways from genotype to phenotype.

Summary

While natural selection acts on phenotypes and our primary observations are on phenotypes, the underlying phenomena are genetic. Our premise is that understanding the basis of phenotypic variation and of adaptive evolution must depend on knowledge of the genetics of the traits of interest. These traits: fitness itself, fitness components or life history traits, and morphological, physiological or behavioural traits affecting fitness, are primarily quantitative. Developing the quantitative genetics of these traits in natural populations will require considerable ecological information, and we stress the intimate relation between ecology and genetics in the study of adaptive evolution.

Following some general comments on genetic variation for quantitative traits, primarily noting that there is ample evidence of genetic variation for fitness components and other traits of interest, the relationship between qualitative and quantitative variation is discussed. The nature of genetic variation for quantitative traits is outlined, and problems of the partitioning of variation in natural populations are noted. Major attention is devoted to the organisation and maintenance of quantitative genetic variation, and we review the theoretical models. It is emphasised that the problem of explaining the maintenance of quantitative genetic variation is not one of a deficiency of possible explanations, but rather of their experimental evaluation and assessment of their relative importance in natural populations. Increased understanding of the connections between genotypes and quantitative trait phenotypes will be a significant factor in understanding adaptive evolution.

References

Anderson WW (1971) Genetic equilibrium and population growth under density-regulated selection. Am Nat 105:489–498

Anderson WW, King CE (1970) Age-specific selection. Proc Natl Acad Sci USA 66:780–786

Atkinson WD (1979) A field investigation of larval competition in domestic *Drosophila*. J Anim Ecol 48:91–102

Baldwin JD, Dingle H (1986) Geographic variation in the effects of temperature on life-history traits in the large milkweed bug *Oncopeltis fasciatus*. Oecologia (Berl) 69:64–71

Barker JSF (1974) The state of information concerning deviations from additivity of gene effects. Proc 1st World Congr Genet Appl Livestock Prod 1:373–383

Barker JSF (1985) Potential contributions of genetics to animal production. Proc 3rd AAAP Anim Sci Congr 1:22–32

Barton NH (1986) The maintenance of polygenic variation through a balance between mutation and stabilizing selection. Genet Res 47:209–216

Bradshaw AD (1984) The importance of evolutionary ideas in ecology–and vice versa. In: Shorrocks B (ed) Evolutionary Ecology. Blackwell, Oxford, pp 1–25

Cade WH (1984) Genetic variation underlying sexual behaviour and reproduction. Am Zool 24: 355–366

Carson HL, Templeton AR (1984) Genetic revolutions in relation to speciation phenomena: the founding of new populations. Annu Rev Ecol Syst 15:97–131

Caspari E (1950) On the selective value of the alleles *Rt* and *rt* in *Ephestia kühniella*. Am Nat 84: 367–380

Chakraborty R, Ryman N (1983) Relationship of mean and variance of genotypic values with heterozygosity per individual in a natural population. Genetics 103:149–152

Charlesworth B (1970) Selection in populations with overlapping generations. I. The use of Malthusian parameters in population genetics. Theor Popul Biol 1:352–370

Charlesworth B (1971) Selection in density-regulated populations. Ecology 52:469–474

Charlesworth B (1980) Evolution in Age-Structured Populations. Cambridge Univ Press, Cambridge

Charlesworth D, Charlesworth B (1975) Theoretical genetics of Batesian mimicry. II. Evolution of supergenes. J Theor Biol 55:305–324

Clare MJ, Luckinbill LS (1985) The effects of gene-environment interaction on the expression of longevity. Heredity 55:19–29

Clarke B (1972) Density-dependent selection. Am Nat 106:1–13

Clarke B (1975) The contribution of ecological genetics to evolutionary theory: detecting the direct effects of natural selection on particular polymorphic loci. Genetics 79:101–113

Clarke B, Whitehead DL (1984) Opportunities for natural selection on DNA and protein at the *Adh* locus in *Drosophila melanogaster*. Dev Genet 4:425–438

Clegg MT, Epperson BK (1985) Recent developments in population genetics. Adv Genet 23: 235–269

Crow JF, Kimura M (1964) The theory of genetic loads. Proc XI Int Congr Genet, pp 495–505

Crow JF, Kimura M (1970) An introduction to population genetics theory. Harper & Row, New York

Darwin C (1859) On the origin of species by means of natural selection, or the preservation of favoured races in the struggle for life. John Murray, London

Dawson PS, Riddle RA (1983) Genetic variation, environmental heterogeneity, and evolutionary stability. In: King CE, Dawson PS (eds) Population biology. Retrospect and prospect. Columbia Univ Press, New York, pp 147–170

Dingle H (1981) Geographic variation and behavioral flexibility in milkweed bug life histories. In: Denno RF, Dingle H (eds) Insects and life history patterns: geographic and habitat variation. Springer, Berlin Heidelberg New York, pp 57–73

Dingle H, Baldwin JD (1983) Geographic variation in life histories: a comparison of tropical and temperate milkweed bugs (*Oncopeltis*). In: Brown VK, Hodek J (eds) Diapause and life cycle strategies. Junk, The Hague, pp 143–166

Dobzhansky Th (1973) Nothing in biology makes sense except in the light of evolution. Am Biol Teach, March 1973, pp 125–129

Falconer DS (1952) The problem of environment and selection. Am Nat 86:293–298

Falconer DS (1981) Introduction to quantitative genetics. Longman, London

Felsenstein J (1986) Population differences in quantitative characters and gene frequencies: a comment on papers by Lewontin and Rogers. Am Nat 127:731–732

Fisher RA (1930) The Genetical Theory of Natural Selection. Clarendon, Oxford

Fleming WH (1979) Equilibrium distributions of continuous polygenic traits. SIAM J Appl Math 36:148–168

Frankham R (1980) Origin of genetic variation in selection lines. In: Robertson A (ed) Selection experiments in laboratory and domestic animals. Commonwealth Agriculture Bureaux, Farnham Royal, UK, pp 56–68

Franklin IR (1982) Population size and the genetic improvement of animals. In: Barker JSF, Hammond K, McClintock AE (eds) Future developments in the genetic improvement of animals. Academic Press Australia, Sydney, pp 181–196

Gillespie JH (1984) Pleiotropic overdominance and the maintenance of genetic variation in polygenic characters. Genetics 107:321–330

Grant BR (1985) Selection on bill characters in a population of Darwin's finches: *Geospiza conirostris* on Isla Genovesa, Galápagos. Evolution 39:523–532

Grant PR, Price TD (1981) Population variation in continuously varying traits as an ecological genetics problem. Am Zool 21:795–811

Gunawan B (1981) The relationship between quantitative and allozyme variation in *Drosophila buzzatii*. In: Gibson JB, Oakeshott JG (eds) Genetic studies of *Drosophila* populations. Proceedings of the Kioloa Conference. Australian National University, Canberra, pp 147–157

Hammond K, James JW (1970) Genes of large effect and the shape of the distribution of a quantitative character. Aust J Biol Sci 23:867–876

Hilbish TJ, Koehn RK (1985) Dominance in physiological phenotypes and fitness at an enzyme locus. Science 229:52–54

Hill WG (1982) Predictions of response to artificial selection from new mutations. Genet Res 40: 255–278

Istock CA (1970) Natural selection in ecologically and genetically defined populations. Behav Sci 15:101–115

Istock C (1984) Boundaries to life history variation and evolution. In: Price PW, Slobodchikoff CN, Gaud WS (eds) A new ecology. Wiley, New York, pp 143–168

Kimura M (1965) A stochastic model concerning the maintenance of genetic variability in quantitative characters. Proc Natl Acad Sci USA 54:731–736

King CE, Anderson WW (1971) Age-specific selection. II. The interaction between r and K during population growth. Am Nat 105:137–156

Koehn RK (1978) Physiology and biochemistry of enzyme variation: the interface of ecology and population genetics. In: Brussard PF (ed) Ecological genetics: the interface. Springer, Berlin Heidelberg New York, pp 51–72

Krimbas CB (1984) On adaptation, Neo-Darwinian tautology and population fitness. Evol Biol 17:1–57

Lande R (1976) The maintenance of genetic variability by mutation in a polygenic character with linked loci. Genet Res 26:221–235

Lande R (1979) Quantitative genetic analysis of multivariate evolution, applied to brain: body size allometry. Evolution 33:402–416

Lande R (1980) The genetic covariance between characters maintained by pleiotropic mutations. Genetics 94:203–215

Lande R (1981) The minimum number of genes contributing to quantitative variation between and within populations. Genetics 99:541–553

Lande R (1982) A quantitative genetic theory of life history evolution. Ecology 63:607–615

Lande R, Arnold SJ (1983) The measurement of selection on correlated characters. Evolution 37:1210–1226

Lawrence MJ (1984) The genetical analysis of ecological traits. In: Shorrocks B (ed) Evolutionary ecology. Blackwell, Oxford, pp 27–63

Lerner IM (1954) Genetic homeostasis. Oliver & Boyd, Edinburgh

Lewontin RC (1974a) The genetic basis of evolutionary change. Columbia Univ Press, New York

Lewontin RC (1974b) The analysis of variance and the analysis of causes. Am J Hum Genet 26: 400–411

Lewontin RC (1984) Detecting population differences in quantitative characters as opposed to gene frequencies. Am Nat 123:115–124

Lewontin RC (1985) Population genetics. Annu Rev Genet 19:81–102

Lewontin RC (1986) A comment on the comments of Rogers and Felsenstein. Am Nat 127: 733–734

Lively CM (1986) Predator-induced shell dimorphism in the acorn barnacle *Chthamalus anisopoma*. Evolution 40:232–242

Livshits G, Kobyliansky E (1985) Lerner's concept of developmental homeostasis and the problem of heterozygosity level in natural populations. Heredity 55:341–353

Lloyd DG (1977) Genetic and phenotypic models of natural selection. J Theor Biol 69:543–560

Loveless MD, Hamrick JL (1984) Ecological determinants of genetic structure in plant populations. Annu Rev Ecol Syst 15:65–95

Luckinbill LS, Clare MJ (1985) Selection for life span in *Drosophila melanogaster*. Heredity 55: 9–18

Luckinbill LS, Arking R, Clare MJ, Cirocco WC, Buck SA (1984) Selection for delayed senescence in *Drosophila melanogaster*. Evolution 38:996–1003

Mackay TFC (1981) Genetic variation in varying environments. Genet Res 37:79–93

Maynard Smith J (1975) The theory of Evolution. Penguin Books, Harmondsworth

Maynard Smith J, Burian R, Kauffman S, Alberch P, Campbell J, Goodwin B, Lande R, Raup D, Wolpert L (1985) Developmental constraints and evolution. Q Rev Biol 60:265–287

Michod RE (1984) Constraints on adaptation, with special reference to social behaviour. In: Price PW, Slobodchikoff CN, Gaud WS (eds) A new ecology. Wiley, New York, pp 253–278

Milkman R (1982) Toward a unified selection theory. In: Milkman R (ed) Perspectives on evolution. Sinauer, Sunderland, Mass, pp 105–118

Mitchell-Olds T, Rutledge JJ (1986) Quantitative genetics in natural plant populations: a review of the theory. Am Nat 127:379–402

Mitton JB, Grant MC (1984) Associations among protein heterozygosity, growth rate, and developmental homeostasis. Annu Rev Ecol Syst 15:479–499

Nei M, Graur D (1984) Extent of protein polymorphism and the neutral mutation theory. Evol Biol 17:73–118

Price TD, Grant PR (1984) Life history traits and natural selection for small body size in a population of Darwin's finches. Evolution 38:483–494

Prout T (1958) A possible difference in genetic variance between wild and laboratory populations. Drosophila Inf Serv 32:148–149

Rendel JM (1967) Canalisation and gene control. Logos, London

Robertson A (1956) The effect of selection against extreme deviants based on deviation or on homozygosis. J Genet 54:236–248

Robertson A (1959) The sampling variance of the genetic correlation coefficient. Biometrics 15: 469–485

Rogers AR (1986) Population differences in quantitative characters as opposed to gene frequencies. Am Nat 127:729–730

Rose MR (1982) Antagonistic pleiotropy, dominance and genetic variation. Heredity 48:63–78

Rose MR (1983) Further models of selection with antagonistic pleiotropy. In: Freedman HI, Strobeck C (eds) Population biology. Springer, Berlin Heidelberg New York, pp 47–53

Rose MR (1984) Laboratory evolution of postponed senescence in *Drosophila melanogaster*. Evolution 38:1004–1010

Rose MR (1985) Life history evolution with antagonistic pleiotropy and overlapping generations. Theor Popul Biol 28:342–358

Rose MR, Charlesworth B (1980) A test of evolutionary theories of senescence. Nature 287: 141–142

Rose MR, Charlesworth B (1981) Genetics of life history in *Drosophila melanogaster*. II. Exploratory selection experiments. Genetics 97:187–196

Roughgarden J (1971) Density-dependent natural selection. Ecology 52:453–468

Sammeta KPV, Levins R (1970) Genetics and ecology. Annu Rev Genet 4:469–488

Service PM, Rose MR (1985) Genetic covariation among life-history components: the effect of novel environments. Evolution 39:943–945

Stearns SC (1984) How much of the phenotype is necessary to understand evolution at the level of the gene? In: Wöhrmann K, Loeschcke V (eds) Population Biology and Evolution. Springer, Berlin Heidelberg New York, pp 31–45

Tachida H, Mukai T (1985) The genetic structure of natural populations of *Drosophila melanogaster*. XIX. Genotype-environment interaction in viability. Genetics 111:43–55

Templeton AR (1981) Mechanisms of speciation – a population genetic approach. Annu Rev Ecol Syst 12:23–48

Templeton AR (1982) Adaptation and the integration of evolutionary forces. In: Milkman R (ed) Perspectives on evolution. Sinauer, Sunderland, Mass, pp 15–31

Turelli M (1984) Heritable genetic variation via mutation-selection balance: Lerch's zeta meets the abdominal bristle. Theor Popul Biol 25:138–193

Turelli M (1985) Effects of pleiotropy on predictions concerning mutation selection balance for polygenic traits. Genetics 111:165–195

Turelli M (1986) Gaussian versus non-Gaussian genetic analyses of polygenic mutation-selection balance. In: Kerlin S, Nevo E (eds) Evolutionary processes and theory. Academic Press, New York, pp 607–628

Van Delden W (1982) The alcohol dehydrogenase polymorphism in *Drosophila melanogaster*. Evol Biol 15:187–222

Via S (1984a) The quantitative genetics of polyphagy in an insect herbivore. I. Genotype-environment interaction in larval performance on different host plant species. Evolution 38:881–895

Via S (1984b) The quantitative genetics of polyphagy in an insect herbivore. II. Genetic correlations in larval performance within and among host plants. Evolution 38:896–905

Via S, Lande R (1985) Genotype-environment interaction and the evolution of phenotypic plasticity. Evolution 39:505–522

Waddington CH (1953) Genetic assimilation of an acquired character. Evolution 7:118–126

Waddington CH (1957) The strategy of the genes. Allen & Unwin, London

Wallace B (1984) Adaptation, Neo-Darwinian tautology, and population fitness: a reply. Evol Biol 17:59–71

Watt WB, Carter PA, Blower SM (1985) Adaptation at specific loci. IV. Differential mating success among glycolytic allozyme genotypes of *Colias* butterflies. Genetics 109:157–175

Wilson AC (1976) Gene regulation in evolution. In: Ayala FJ (ed) Molecular evolution. Sinauer, Sunderland, Mass, pp 225–236

Wright S (1931) Evolution in Mendelian populations. Genetics 16:97–159

Wright S (1932) The roles of mutation, inbreeding, crossbreeding and selection in evolution. Proc VI Int Congr Genet 1:356–366

Wright S (1969) Evolution and the genetics of populations, vol 2. The theory of gene frequencies. The University of Chicago Press, Chicago

Yoo BH (1980a) Long-term selection for a quantitative character in large replicate populations of *Drosophila melanogaster*. I. Response to selection. Genet Res 35:1–17

Yoo BH (1980b) Long-term selection for a quantitative character in large replicate populations of *Drosophila melanogaster*, part 3. The nature of residual genetic variability. Theor Appl Genet 57:25–32

Zink RM, Smith MF, Patton JL (1985) Associations between heterozygosity and morphological variance. J Hered 76:415–420

Chapter 2 Genetic Correlations: The Quantitative Genetics of Evolutionary Constraints

A. G. CLARK[1]

Dictionary definitions of "constraint" generally indicate some aspect of restricting or confining the possible states or actions of individuals or systems. In the disciplines of linear and dynamic programming, a constraint is a function or inequality specifying the range of permissible values of variables. If the constraints were removed, the dynamical and equilibrium properties of a system of equations may be quite altered. A constraint may be either fixed, specifying an absolute limit on a particular variable, or it may be dynamic and restrict the joint behavior of two or more variables. For the purpose of this discussion, a genetic constraint will be defined functionally as those aspects of the inheritance of traits that prevent natural selection from resulting in a steepest ascent approach of the mean phenotype to an optimum. Genetic constraints can be fixed, as in the case when genetic variation for a particular phenotype does not exist, or dynamic, as in the case of genetic correlations depending on allele frequencies and linkage disequilibria. The primary genetic reason to be suspicious of the adaptationist program (Gould and Lewontin 1979) is the prevalence of genetic correlation. The intent of this chapter is to examine constraints as themselves being evolved traits, just as the processes of Mendelian genetics are themselves the products of evolution. The utility of specifying an explicit relationship between phenotypic characters and fitness based on demographic or physiological principles will be demonstrated.

With a single genetic locus determining a character, classical population genetics theory gives explicit predictions of the outcome of natural selection. In particular, with heterozygote viability advantage, allele frequencies will change monotonically until the population mean fitness is maximized. This statement of the fundamental theorem of natural selection (Fisher 1958) is known to be violated by a number of phenomena relating to the genetic transmission or the mode of selection (Ewens 1979). If one allele is meiotically driven, for example, mean fitness may no longer be maximized. This is perhaps the simplest example of "antagonistic pleiotropy"; if an allele has an effect on more than one component of its transmission, then maximization of mean fitness may fail.

The fundamental theorem can also fail due to recombination. Analysis of two locus models have led to a clear delineation of the conditions when selection on a pair of loci will not maximize fitness. In particular, even in the absence of multiplicative epistasis, mean fitness may fail to be maximized (for review, see Karlin 1975). A typ-

---

[1] Department of Biology, Pennsylvania State University, University Park, PA 16802, USA

Genetic Constraints on Adaptive Evolution
Ed. by V. Loeschcke
© Springer-Verlag Berlin Heidelberg 1987

ical quantitative genetic study will manifest the effects of this type of linkage or pleiotropy as a genetic correlation. A simple sib analysis or artificial selection experiment will not reveal whether the genetic correlation is due to linkage or pleiotropy, but the consequences of these two mechanistically distinct genetic constraints could be quite different. The quantitative genetic theory dealing with the joint evolution of a pair of traits with genetic correlations generally assumes a constant variance-covariance matrix, and the stability of the variance-covariance matrix is likely to depend on the underlying mechanism of the genetic correlation.

Before reviewing the quantitative genetic theory of selection on two characters, it should be made clear that there are other sources of genetic constraint besides genetic correlation. Perhaps the simplest conceptually is that an absence of genetic variation would constrain the evolution of a population. Antonovics (1976) argues that this is not an important constraint for many ecologically significant characters, simply because there is direct evidence for additive genetic variance. Another constraint that is particularly difficult to measure by quantitative genetic techniques is epistatic interaction among loci. Although there are methods to estimate epistatic variance, they are rather elaborate, and require large and complex sampling efforts. Even though epistasis may be quite important in the mapping from genotypes to phenotypes (Lewontin 1974; Barker 1979), we will begin by assuming interlocus additivity, and concentrate on the easily measured constraint caused by genetic correlation. A partial justification for this assumption is that it is the additive genetic variance that determines the response to selection. For background we first briefly review the approach of Lande (1979, 1980, 1984) in developing the recurrence system for multivariate quantitative characters in terms of linear transformations.

Selection on Two Characters

Some rather restrictive assumptions must be made in developing the quantitative genetic theory of multivariate phenotypic evolution, but the power of the conclusions in generating hypotheses makes the approach particularly attractive. Perhaps the most important assumption is that an appropriate scaling can be found such that there is a multivariate Gaussian distribution of phenotypes. If we assume that many loci are involved in determining the traits, and that the effects of these loci are additive, then the Central Limit Theorem applies in summing the allelic effects to yield a Gaussian distribution.

Ideally, this scaling also renders means and variances independent, and stabilizes variances. In the concluding remarks, we will briefly discuss the likely limitations that the theory will suffer as a result of these assumptions.

Consider two phenotypic traits, z_1 and z_2 that show polygenic inheritance. Assuming no sexual dimorphism, the genetic variance-covariance matrix can be written:

$$G = \begin{bmatrix} V_{z1} & C_{z1,z2} \\ C_{z1,z2} & V_{z2} \end{bmatrix}, \tag{1}$$

where V_{z1} and V_{z2} are the additive genetic variances of traits z_1 and z_2, respectively, and $C_{z1,z2}$ is the covariance of additive genetic effects. Define $W(z_1,z_2)$ as the function describing the fitness of an individual with phenotype (z_1,z_2). The formulation for bivariate quantitative genetic selection response follows directly from the univariate formulation, and derives a linear transformation of phenotypic means assuming weak selection and additivity of effects. The bivariate and multivariate formulations need to account for the fact that selection on any one character may influence the mean phenotype of another character through correlated selection responses if genetic covariances are non-zero. Following Lande (1979), the vector describing the change in the mean phenotypes ($\Delta\bar{z}$) is

$$\Delta\bar{z} = G \nabla \ln\bar{W}, \tag{2}$$

where $\nabla \ln\bar{W}$ is the selection gradient. This linearization is approximately correct if the distribution of phenotypic and additive genetic effects are Gaussian. In this case

$$\nabla \ln\bar{W} = \begin{bmatrix} \partial \ln\bar{W}/\partial\bar{z}_1 \\ \partial \ln\bar{W}/\partial\bar{z}_2 \end{bmatrix}. \tag{3}$$

The selection gradient specifies the direction of steepest ascent up the fitness surface. If G is the identity matrix (i.e., all genetic covariances are zero), then natural selection drives the mean phenotype up the gradient. With non-zero genetic covariances, the result of natural selection may be to force the vector of mean phenotypic change away from the gradient. Genetic covariances can in fact result in phenotypic changes in a direction opposing the direction of the direct selective effect on a trait, a phenomenon called "antagonistic selection" (Lande 1979). For this reason, genetic covariances are critical to the issue of genetic constraints.

The above models do not incorporate explicit genetic transmission, but rather assume that many loci are involved, with approximately equal effect on the traits of interest. Numerical studies have been applied that have an underlying multilocus Mendelian transmission (see below), and they help to determine the range of parameter values under which the approximations of the quantitative genetic theory hold. Gimelfarb (1986) presents a model with two linked genetic loci each having pleiotropic effects on two traits. One locus has a similar effect on both traits, while the other locus has an antagonistic effect on the two characters. After establishing a correspondence between the traits and fitness (with zero multiplicative epistasis), Gimelfarb (1986) uses this simple model to illustrate some remarkably unfortunate things. The two additive characters can maintain variation (an "internal" equilibrium in the two-locus model) despite the stabilizing selection acting directly on those characters. If one were to attempt to measure components of genetic variance and covariances for these traits, one would find ample additive genetic variance but no genetic covariance, despite the pleiotropic relations of the two loci. One would be faced with having to explain a large amount of additive genetic variance in the face of stabilizing selection and no genetic correlation. As Gimelfarb (1986) shows, it would be almost impossible to clarify the underlying effects of the two loci using classical quantitative genetics methods.

Much of the theory that has followed from Lande's approach is based in part on the assumption of constancy of the genetic variance-covariance matrix. This assumption often seems empirically reasonable: many selection experiments have shown fairly constant response over many generations. For applications such as the measurement of natural selection (Lande and Arnold 1983), evolutionary changes in genetic variances and covariances are not really relevant. However, over evolutionary time there may be considerable opportunity for the genetic covariance matrix to itself be molded by the processes of mutation, drift, and selection (Lofsvold 1986). Lande (1980), Bulmer (1980), and Turelli (1985, 1986) considered the evolution of genetic covariances in a model with Gaussian selection and mutation, but before we consider those results, let us return to the univariate problem.

Mutation-Selection Balance for a Single Character

In order to examine the evolution of the genetic covariance matrix, it is necessary to consider the joint effects of mutation and selection. Other forces that may be involved (but that we will not be concerned with here) include population structure and genetic drift. The quantitative genetic theory of mutation-selection balance has been thoroughly reviewed by Turelli (1984, 1985, 1986), who also extended the results under slightly different assumptions. Elsewhere in this volume (see Barker and Thomas), the assumptions of the Gaussian "Kimura-Lande-Fleming" model are presented in detail, as are some of the limitations of the model. Our purpose here is to give the explicit formulation of the KLF model, in order to set the foundation for exploring the mutation-selection model with bivariate phenotypes.

The objective of the univariate model is to derive the distribution of allelic effects in a population where mutation tends to increase the variance and balancing selection tends to reduce the variance. In simple intuitive terms, if the fitness of individuals in the tails of the phenotypic distribution is arbitrarily low, then the forces of mutation and selection must come to some balance where the variance in the distribution of allelic effects is constrained from growing without bound. The subtleties enter these models when one tries to explain the amount of genetic variance actually seen in natural populations in terms of rough estimates of mutation rates, the distribution of mutation effects, and the distribution of fitnesses. Controversy arises because there is a continuum of mutation rates and magnitudes of mutational effects and fitnesses that result in a given equilibrium genetic variance. For some of these parameters, the equilibrium distribution of allelic effects will be very nearly Gaussian, and in other cases it may deviate strongly from a Gaussian distribution. Different portions of the parameter space require different approximations to be able to handle the recursion of allelic effects analytically.

Following Kimura (1965), Lande (1975), and Turelli (1984), let $p_t(x)$ represent the distribution of allelic effects before selection at generation t. In each generation, selection takes place, then mutation occurs, and finally the population is sampled. This haploid model is an approximation to a diploid model with weak selection, and with diploids the sequence of events of each generation is zygote formation, selection, meiosis, and mutation, then population censusing. Define $p'_t(x)$ as the density of allelic effects after selection, and μ as the mutation rate per generation. Let $g(x)$ be the density function for the phenotypic effect of a mutation. The recurrence equation for the distribution of allelic effects is

$$p_{t+1}(x) = (1-\mu)p'_t(x) + \mu\int p'_t(y)g(x-y)dy. \tag{4}$$

Selection can be described by the function $w(x)$, giving the mean fitness of phenotype x. The distribution of allelic effects after selection is

$$p'_t(x) = \frac{w(x)p_t(x)}{\int w(y)p_t(y)dy}. \tag{5}$$

Selection operates by weighting phenotypes by their respective fitness, and frequencies are normalized to sum to 1. Kimura (1965) examined a one-locus continuum-of-alleles model with mutational effects having a Gaussian distribution with small variance. Using a quadratic selection function, the equilibrium distribution of allelic effects is approximately Gaussian with variance:

$$V_g = \sqrt{\mu m^2 V_s}, \tag{6}$$

where V_s is the variance in the fitness distribution and m^2 is the variance in mutational effects. Kimura's (1965) analysis is based on a continuous time model, and Latter (1970) similarly applied a Gaussian approximation to a discrete-time version of the one-locus mutation-selection model. Lande (1975), Fleming (1979), and Nagylaki (1984) extended the analysis to consider many linked loci. Lande (1975) and Bulmer (1980) specifically considered linkage disequilibrium, and found that the equilibrium genetic variance was only slightly influenced by linkage disequilibrium. Fleming (1979) obtained an expression for the equilibrium genetic variance whose first term is given in Eq. (6). Numerical analyses by Turelli (1984) show the prediction of Fleming (1979), incorporating a correction for departure from normality, to be quite accurate. The Gaussian prediction was found to be acceptable only with a very high mutation rate and weak selection.

All of these models are based on either the one-locus Eq. (4) or its multilocus generalization. They differ only in the way approximations are made to find solutions. The distribution of allelic effects is in fact a mixture distribution with complex properties, necessitating an approximate solution. Turelli (1984) questions the validity of the Gaussian approximation. Lande (1975) saw that the distribution of allelic effects would be approximately Gaussian if $m_i^2 << V_g$, and Turelli (1984) pointed out that after Eq. (6) is applied, this becomes:

$$\mu_i >> m_i^2/V_s. \tag{7}$$

Turelli (1984) argues that the data support neither of these constraints, since they imply unrealistically high mutation rates. Resolution of the utility of the approximation will depend on experimental results addressing whether these constraints are acceptable. Turelli (1984), in particular, argues that the effects of a mutation (m_i) are often larger than stated by this approximation, and the rates of mutation are much lower. The "house-of-cards" approximation of Kingman (1978) allows the effects of new mutations to overwhelm the existing phenotype, so that a new mutant's phenotype is independent of its phenotypic state before the mutation. The house-of-cards model provides an acceptable approximation to the equilibrium of the KLF model under conditions of the parameters when the direction of the inequality in Eq. (7) is reversed. This leads to the equation for the distribution of allelic effects when the population is near equilibrium:

$$P_{t+1}(x) = \frac{(1\text{-}\mu)p_t(x)w(x)}{\overline{w}_t} + \mu g(x). \tag{8}$$

The first term on the right is the distribution of non-mutants after selection, and $g(x)$ is the distribution of allelic effects of mutants. It is clear that $g(x)$ is independent of the premutation state.

After a number of approximations, Turelli (1984) derives the result that the equilibrium genetic variance in the n-locus house-of-cards model is

$$V_g = 4n\mu V_s. \tag{9}$$

This result is also obtained for the n-diallelic locus approximation to the Gaussian model (Latter 1960; Bulmer 1972), and in both cases it is assumed that the n loci are equally mutable and are in global linkage equilibrium. This result is not entirely intuitive because of the absence of a dependence on m, the magnitude of effects of mutation. Once m is large enough to swamp the current genetic variance, the equilibrium genetic variance becomes independent of m. Turelli (1984) presents a number of numerical simulations of diallelic and triallelic loci, and generally finds that results are sensitive to the magnitudes of the parameters. With reference parameters $\mu=10^{-4}$, $m^2=0.05$, and $V_s=20$, Turelli (1984) finds that the Gaussian approximation is best for higher mutation rates, lower variance in mutation effects, and weaker selection. The house-of-cards approximation is better for low mutation rates, large mutation effects, and stronger selection. Although Turelli (1984) presents a strong case that the Gaussian approximation requires unrealistic parameters, more data would be useful to determine what are biologically reasonable parameters. It is premature to conclude that the theory of mutation-selection balance adequately explains the currently observed levels of additive genetic variance. As we will see below, even if we had complete confidence in the parameter estimates, it would be very difficult to test the hypothesis that the observed genetic variance is explained by mutation-selection balance, due to the influence of selection on correlated characters.

The extent of the problem is underscored by Barton (1986), who finds a multiplicity of equilibria under diallelic mutation-selection balance models. Both the Kimura-Lande-Fleming model and the house-of-cards model give the same predicted equilibrium additive genetic variance in the diallelic case, and both assume that the equilibrium mean phenotype is at the evolutionary optimum. Although Lande (1979) showed that mean fitness is maximized in a multivariate phenotypic selection model with no mutation, Barton (1986) showed that, whichever approximation is taken, there may be multiple equilibria in the diallelic mutation-selection balance model. Barton's (1986) conclusion may be even more sobering than Turelli's, in that it shows that the equilibrium genetic variance cannot be predicted by a model incorporating mutation rates and distributions of mutation and selection effects, because the equilibrium variance will depend on the evolutionary history (the "initial conditions") of the population. The generality of this result is not entirely clear, however, and it may depend on the symmetry that is built into the model.

Despite the difficulty these models face in explaining the observed additive genetic variance by mutation-selection balance, they do provide a formal structure for extended models that allow mutation and selection in more than one character. Addition of phenotypic characters opens the possibility of other modes of maintaining genetic variance, particularly through selection on alleles with pleiotropic effects. Unfortunately, the complete verification of these models by experimentation and by estimating the relevant parameters seems unpromising. Verification would require good estimates of bivariate distributions of mutations and selective effects, a task that is surely more onerous than univariate estimates. In the next section the model of pleiotropic mutation by Lande (1980) is presented. This model derives conditions for the evolution of genetic covariances when linkage is sufficiently loose, and the distributions of effects of mutations and selection are multivariate Gaussian.

Mutation-Selection Balance for More Than One Character

Given that the problem of mutation-selection balance for one character leads to a variety of conflicting interpretations, it may seem ambitious to attack the problem of mutation-selection balance for two or more characters. Lande (1980) presents a theoretical analysis for the evolution of correlated characters using the Gaussian assumptions and linearization, and since this is precisely the model that describes the evolution of genetic correlations, it is suitable to examine in some more detail. In the last two sections of this chapter, specific bivariate selection functions will be derived based on biological principles of the phenotypes in question. In these cases, the models provide a focus for generating hypotheses about the maintenance of additive variance and covariance for the particular characters.

As in the model for multivariate selection presented above, it is assumed that environmental and genetic effects are additive on some scale. The approach is to assume that the multivariate distributions of phenotypes, additive genetic effects, fitnesses, and mutational effects are multivariate Gaussian, and that they are specified by their respective variance-covariance matrices. All of the recurrences that are derived are in terms of these covariance matrices, and hence are linearizations of the dynamic processes. Linearization in this manner is valid for weak selection and small mutational effects.

If the fitness of an individual with a vector of phenotypic measures z is multivariate Gaussian and $\hat{z}$ is the phenotypic optimum it can be written:

$$W(z) = \exp\left\{-.5(z-\hat{z})W^{-1}(z-\hat{z})^T\right\}, \tag{10}$$

and Lande (1980) shows with a nice analogy to multiple regression that the covariance matrix of the phenotypic distribution after selection (P') is:

$$P' = P-P(W+P)^{-1}P. \tag{11}$$

Again, recall the implicit assumption that selection is weak. Mutation perturbs the genetic covariance matrix in a way that can be written as a simple sum of the premutation genetic covariance matrix and a covariance matrix of mutational effects. There is here an implicit assumption that the distribution of allelic effects will remain multivariate Gaussian, and this is an acceptable approximation if mutational effects are sufficiently small. The recurrence equations of the covariance matrix of allelic effects incorporate both recombination and pleiotropic effects. Lande (1980) derives several observations from this system of equations. The first is that the time course for changes in genetic covariances is on the same scale as the time course of changes in genetic variances. Under similar assumptions as the univariate Gaussian model, Lande (1980) derives the equilibrium variance-covariance matrix of effects of locus i as

$$G_i = W^{\frac{1}{2}}(\mu_i W^{-\frac{1}{2}} U_i W^{-\frac{1}{2}})^{\frac{1}{2}} W^{\frac{1}{2}} \tag{12}$$

where W is the covariance matrix of fitnesses and U_i is the covariance matrix of mutational effects. If W and U_i are diagonal matrices, so that there are no correlations in mutational or selective effects, then the equilibrium variances predicted by Eq. (12) agree with those of Eq. (6). In this case, the characters are evolving independently. A primary result when there are correlated effects in mutation and selection was succinctly stated by Lande (1980, p. 216): "The genetic covariance matrix of the characters is then determined by a balance between the patterns of pleiotropic mutations and multivariate selection on the phenotype".

The approximation of a multivariate Gaussian distribution of allelic effects is possible under the assumption of weak selection and small mutational effects (Lande 1980). Just as these assumptions were questioned and replaced by a univariate house-of-cards model, low mutation rates and greater mutational effects can be accomodated by a multivariate house-of-cards model (Turelli 1985). Analytical and numerical studies of Turelli (1985) show that (1) increasing intensity of selection on unidentified correlated characters causes reduced equilibrium genetic variance in the house-of-cards model, (2) in the absence of correlated mutational and selective effects the Gaussian model indicates that traits evolve independently, while the house-of-cards model indicates that equilibrium variances are still interdependent, and (3) despite large differences in the equilibrium genetic variance between the Gaussian and house-of-cards models, the equilibrium genetic correlations agree very well.

For the numerical simulations that will be discussed in the next section, it is not necessary to assume that the effects of mutations are so small that the recurrences can be treated as linear operators. Similarly, we are not restricted to Gaussian fitness functions, nor must we make the approximation that the distribution of allelic effects will be Gaussian. Numerical simulations permit an examination of the consequences of the parameter space in the continuum from the valid region for the Gaussian model to the house-of-cards model (Turelli 1984, 1985). While Turelli (1985) examined the effects of pleiotropic mutations on the maintenance of genetic variance, our primary interest here is to consider the way that mutation and selection act to generate genetic correlations. If the bivariate distribution of allelic effects before selection is written as $p(x,y)$, then the distribution after selection is obtained from the convolution

$$p'(x,y) = \frac{w(x,y)p(x,y)}{\int\int w(u,v)p(u,v)\ du\ dv}. \tag{13}$$

Even if $p(x,y)$ is Gaussian, $p'(x,y)$ may not be (except in the limit of small selective effects). Mutations also enter the population following a bivariate distribution of effects on the two traits. Suppose that selection takes place in juveniles and adults and then mutations occur during the formation of gametes; $p(x,y)$ then describes the distribution of allelic effects in gametes and zygotes. Define $g(x,y)$ as the density of the mutational effects. The recursion is then

$$p_{t+1}(x,y) = (1-\mu)p'_t(x,y) + \mu\int\int p'_t(u,v)g(x-u,y-v)\ du\ dv. \tag{14}$$

As in the univariate KLF model, a fraction $1-\mu$ of the population does not mutate, and of these, a group described by $p'_t(x,y)$ survives to contribute to the next generation. Of the fraction μ that mutates, individuals will have phenotypes (x,y) if they started with phenotype (u,v), survived selection, and mutated from (u,v) to (x,y). Stating the model in this way, it is clear that the distribution of allelic effects, which is described as the genetic covariance matrix in a Gaussian setting, is a mixture distribution of bivariate distributions.

   Although approximations can be made to obtain a bivariate Gaussian distribution from this, Turelli (1986) provides compelling reasons to attempt analyses without a Gaussian assumption. Numerical simulations provide an opportunity to examine the assumptions and obtain results in the region of the parameter space where no analytical expression for the equilibrium distribution may exist. However, a complete understanding of the model can only come from a full analytical treatment, so simulations should be considered as either exploratory or as verification of analytical results.

Numerical Simulations of Bivariate Gaussian Selection

There are two primary purposes in performing numerical simulations of quantitative genetic models. The first is to obtain a description of the equilibrium behavior of the phenotypic distributions predicted by the recurrence equation for the single locus haploid model. The motivation here is to determine the robustness of the assumptions that had to be made to obtain a Gaussian equilibrium distribution. In particular, the recursion indicates that the equilibrium distribution, if it exists, will be a mixture distribution. The mathematics of mixture distributions can become very cumbersome, so theoreticians, beginning with Kimura (1965) have tried to find approximations that enable one to employ simple normal distributions. The second type of simulation specifies an underlying genetic structure, and iterates the Mendelian transmission of these genes in order to verify that the equilibrium distributions of phenotypes are predicted by the underlying multilocus genetic structure. In essence, this tests how well the dynamics of a few segregating genes will be predicted by a continuum-of-alleles model with small additive effects. Turelli (1985) did the latter form of simulation to explore the differences between the Gaussian and house-of-cards approximations. His primary interest was the effect of pleiotropy on the maintenance of genetic variance under conditions of mutation-selection balance. Our focus here is to examine the genetic correlations that result from the same models of mutation-selection balance.

The simulations were performed for the simplest model, namely a one-locus haploid continuum-of-alleles model. First, a bivariate Gaussian distribution of phenotype frequencies was generated, where each individual has measured characters X and Y. A fraction μ of the population mutates each generation, and those that mutate are perturbed in both X and Y by an amount specified by a random variable drawn from a bivariate Gaussian distribution whose means, variances, and correlation are specified. The correlation in this distribution will be referred to as the "mutation correlation". Selection occurs by weighting the probability of survival of each individual by its fitness, where the fitness function can be either Gaussian, directional, or specified by a particular functional model. In the case of Gaussian stabilizing selection, we arbitrarily set the modal fitness at (0,0) with unit variances and leave the correlation to be varied. The correlation in this distribution is called the "selection correlation".

Each generation the distribution of phenotypes is split into a mutated and non-mutated portion, those that mutate are perturbed as described above, and selection operates by removing some individuals. In the recurrence system, selection operates by a convolution of the fitness function with the mutation distribution. In the simulations, the fitness of each individual is calculated from the bivariate Gaussian distribution, and a random number on the interval [0,1] is generated. If the fitness exceeds the random number, the individual survives, if not it dies. This process is continued (sampling with replacement) until the 1000 individuals are selected for the next generation. The sequence of mutation and selection is iterated until the means and variances stabilize (typically about 1000 generations). Stationary distributions were obtained relatively quickly because of the high mutation rates and strong selection employed. Implicit in these simulations is a haploid clonal reproduction, with a reproductive advantage scaled by the fitnesses. This procedure introduces a sampling effect, analogous to genetic drift, but selection can be made strong enough to negate the importance of sampling error. For descriptive purposes, this method provides a simple and quick means to obtain an equilibrium distribution of Eq. (14). Another numerical method for solving the equilibrium distribution of allelic effects is to iteratively perform the numerical integration of Eq. (14). Great care must be taken in this case, because of the compounding round-off errors that would be introduced by the iteration.

The runs with a Gaussian fitness function were done on a grid of mutation and selection correlations, over all values between -0.5 and +0.5 with intervals of 0.1. Figure 1 displays the results of simulations using parameters that were chosen to be in the realm of the Gaussian approximation ($\mu >> m_i^2/V_s$). When the mutation rate is low and the effect of mutations is large ($\mu << m_i^2/V_s$), the house-of-cards approximation is more appropriate. Figure 2 shows the pattern of equilibrium genetic correla-

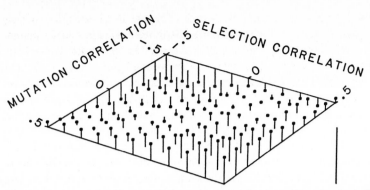

Fig. 1. Results of numerical simulations of the bivariate mutation-selection balance model. Axes are mutation correlation and selection correlation. The "vertical" axis is the equilibrium correlation in allelic effects. The *bar* to the right of the figure represents a correlation of 1. The correlation is indicated by the *dotted end of each stick*, and the *non-dotted end of each stick* lies on the plane (zero correlation). For all 121 simulations represented, the mutation rate $\mu=0.05$, the variance in mutational effects for both characters $m_i^2=0.05$, and the variance in the selection function for each character $V_s=2$. This represents a parameter set typical of the KLF approximation

tion obtained by simulation in the house-of-cards portion of the parameter space. Note that with zero-selection correlation, the equilibrium phenotypic distribution can have non-zero correlation as a consequence of mutation correlation. The converse is also true: even if mutations enter the population in an uncorrelated fashion, natural selection filters what survives and the equilibrium distribution may be correlated. It should be emphasized that an extension of Turelli's (1984, 1985, 1986) arguments to this bivariate case would show that it would be next to impossible to parameterize a simulation of this sort to biologically measured terms. The purpose of these simulations is not to determine whether observed levels of genetic variance and correlation can be explained by mutation-selection balance (although that is a salient

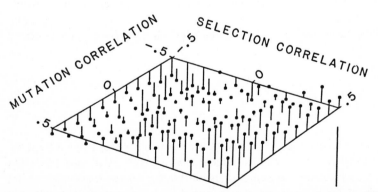

Fig. 2. Results of numerical simulations plotted as in Fig. 1, except here $\mu=0.01$, $m_i^2=0.1$, and $V_s=2$. These are parameters more typical of the house-of-cards approximation, but scaled to a high mutation rate to accelerate the approach to equilibrium

problem). The important point is that observation of genetic correlation does not by itself indicate whether the mutation or selection distributions are correlated. Simulations show that the equilibrium phenotypic distribution can have either the same or reversed correlation from the selection correlation depending on the direction and strength of the mutational correlation and the mutation rate.

The inference of pleiotropic connections between traits from genetic correlations is made even more tenuous when one considers a third trait. It may be quite possible that the two initial characters are each pleiotropically related to the third, but are relatively independent of each other in action. Since it can never be known whether all of the relevant traits have been scored, this problem of hidden characters can never be completely excluded. A hidden trait or mechanism may be invoked in almost any genetic experiment, but in the case of quantitative characters, it seems less legitimate to apply Okkam's razor.

Models that predict directional selection are often of interest, and in these cases the analysis described above should lead to fixation of alleles and contraction of phenotypic variance to some edge of the phenotypic simplex. In these cases we are interested in either the transient behavior of the genetic covariance matrix or in a balance between directional mutation and directional selection. A simple conclusion is that transient genetic correlation is generated if the fitness surface shows particular directions of curvature with respect to the two characters. In mathematical terms, if both

$$\partial^2 W(x,y)/\partial y^2 < 0 \quad \text{and} \quad \partial^2 W(x,y)/\partial x^2 < 0, \tag{15}$$

then the surface is concave down and there is a tendency toward stabilizing the variances. If, in addition, $\partial^2 W(x,y)/\partial x \partial y < 0$, then there will be a tendency to generate positive correlations, and if this inequality is reversed, negative correlations will be generated (implicit in Lande and Arnold 1983). With directional mutation, genetic correlations along the direction of the mean mutational effects will be generated, but these can be opposed by the selection function. Many selection functions that are based on an explicit physiological, demographic, or biochemical model are distinctly non-Gaussian, and may even be more accurately envisaged as directional rather than stabilizing. In the next two sections, examples of models that give an explicit relationship between attributes of two characters and fitness will be examined, with particular emphasis on the origin and maintenance of genetic correlation between these traits.

Application to Life History

A specific problem in population biology that invokes the theory of multivariate phenotypic selection is the evolution of senescence (see Rose et al., and references therein, this Vol.). One of the leading theories for the evolution of senescence invokes "antagonistic pleiotropy", such that alleles that decrease survivorship of older age classes can increase in frequency when rare, provided they have a pleiotropic effect boosting early fitness (generally fecundity at younger ages). An important question is how natural selection shapes the genetic correlations between early and late age traits associated with fitness. The simplest argument is that selection fixes alleles whose effects are positive (increase fitness) on both early and late fitness components. Alleles with deleterious effects on both fitness components are lost, and only those alleles having antagonistically pleiotropic effects will remain segregating in the population. It may not be appropriate to dichotomize allelic effects in this way. An allele may have pleiotropic effects that change the age-specific distribution of fecundity and survivorhsip in such a way that early fecundity and longevity are both augmented, but late fecundity is drastically reduced. Since fitness can be interpreted as the intrinsic rate of increase (Charlesworth 1980) such an allele may have little net effect on fitness, and hence may continue to segregate in a population.

Lande (1982) examined the theoretical quantitative genetics of life history evolution, and defined fitness as a function of age-specific fecundity and mortality in the manner of Charlesworth (1974, 1980). He found that selection maximized the population's intrinsic rate of increase, and that the maximum depended on the genetic correlation among age-specific traits. Here we examine a similar model using an explicit formulation of fitness in terms of the two characters, (1) mean age of reproduction and (2) mean age of death. The question becomes whether the nature of the selection function can generate genetic correlations between these two characters under various conditions of the mutation distribution. If not, then we conclude that selection on life history does not per se generate the genetic correlations seen.

The selection function for life history has been widely studied, and classically fitness is related to the intrinsic rate of increase of a population. If a genotype confers a particular age-specific schedule of survivorship (l_x) and fecundity (m_x), then the intrinsic rate of increase of a population composed exclusively of that genotype is obtained from the Euler equation:

$$1 = \int e^{-rx} l_x m_x \, dx. \tag{16}$$

A convenient way to envisage the age-specific survivorships and fecundities is to use parametric functions. Fecundity schedules often show a unimodal distribution, and a Gaussian distribution on a logarithmic age scale fits many experimental data sets quite well (Fig. 3). Survivorship curves can be thought of as a dose-response relation, where the dose is age in units of time, and the response is death. Dose-response experiments in toxicology are frequently fitted using a statistical technique called probit analysis, which essentially fits a cumulative normal distribution to the data. This procedure can also give very good results with *Drosophila* data (see Fig. 4). In my laboratory, we have observed statistically significant variation among isogenic lines of *Drosophila* in both the mean and variance of the cumulative normal mortality distribution (Hughes and Clark 1987). For simplicity we will describe the age-specific fecundities and survivorships by a single parameter each, namely the mean age of reproduction and the mean time of death. Implicit in this is the assumption that the total lifetime fecundity does not vary among genotypes. Although this is clearly violated by the data, it serves as a good starting point for the analyses because we are not confounding net fecundity differences with purely age-specific effects. If we arbitrarily fix the variances of these two distributions, then we can plot the fitness, obtained from the Euler equation, for any combination of age-specific fecundity and survivor-

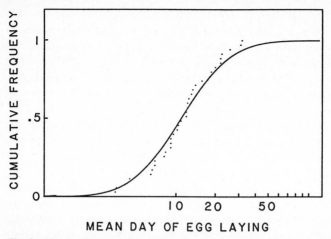

Fig. 3. Age-specific distribution of egg laying in the R6B isogenic line of *Drosophila melanogaster*. *Points* represent the mean day of egg laying (weighted average of days x eggs laid) for single females that were followed throughout their lives and scored for daily egg production. Data were fitted using probit techniques (SAS procedure PROBIT)

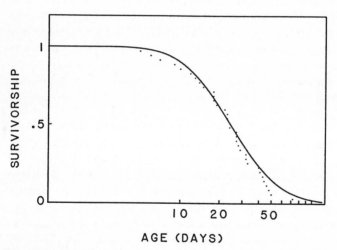

Fig. 4. Survivorship curve for the same flies analyzed in Fig. 3. There was a good fit to the cumulative normal distribution when age was log-transformed

ship. Let D be the mean age of death, and R be the mean age of reproduction. We can specify s/R,D)=e^r as the fitness of such a life history by solving:

$$1 = \int e^{-rx} \phi(x) (1 - \phi(x)) \, dx, \tag{17}$$

where $\phi(x)$ is the Gaussian fecundity-density function with mean R and variance V_r and $\bar{\Phi}(x)$ is the cumulative normal mortality distribution with mean D and variance

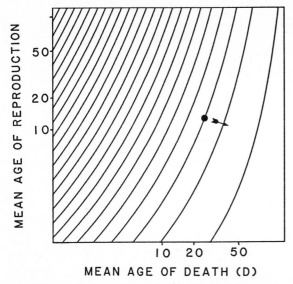

Fig. 5. Fitness surface obtained by applying Euler's equation to a grid of points (mean age of reproduction R, and mean age of death D). For each point we obtain the intrinsic rate of population growth at stable age distribution. *Contours* represent level curves of the polynomial fit to these points. The *point in the interior* represents the line R6B, plotted in Figs. 3 and 4, and the *arrow* indicates the gradient vector from this point. Note that the fitness surface becomes flatter for larger D, representing a decreased strength of selection with increasing age

V_d. V_r and V_d can be thought of as environmental variances, assumed to be constant and independent of genotype. This fully specifies a selection function analogous to Wright's (1969) adaptive topography. In this case the axes are phenotypic characters and the trajectory traced on this topography represents phenotypic distributions, while Wright's adaptive topography represents gene frequency dynamics. It is readily apparent that this function depicts directional selection (Fig. 5), tending to push reproduction as early as possible and delay mortality as long as possible. This is biologically not attainable, so the system may be constrained by the physiologically possible mutations. A population may be kept from attaining the fitness maximum due to mutational correlation or due to directional mutation. It is quite reasonable to invoke the latter, since most mutations are deleterious.

Table 1 shows some results of simulations with a variety of mutation correlations and directional mutation effects. The selection function is specified by Eq. (17) using data in Figs. 3 and 4. From these simulations we can conclude: (1) In the absence of directional mutation, the population would go to fixation for an infinite life span and instantaneous reproduction. (2) Without directional mutation, the transient correlation in allelic effects is positive (but the sign convention is reversed here so this corresponds to a negative correlation with respect to fitness). (3) Positive mutation correlation results in selection for earlier death than zero correlation, but (4) in the absence of mutation correlation, the mean age of death always increases. Thus, despite the fact that genetic correlation is generated by this fitness surface, the model

Table 1. Numerical simulations of a balance between mutation and life history selection

Mutation parameters			===⇒	Equilibrium distribution of allelic effects[a]				
$M_{\mu R}$	$M_{\mu D}$	$V_{\mu R}=V_{\mu D}$	P_μ	R	D	V_r	V_D	ρ_g
0	0	0.01	-0.5	0.26	3.49	0.23	0.33	0.04
			0	0.31	3.20	0.26	0.33	0.49
			0.5	0.54	2.41	0.40	0.46	0.83
		0.05	-0.5	0.31	5.96	0.31	0.77	-0.13
			0	0.42	5.28	0.43	0.66	0.21
			0.5	0.44	4.19	0.40	0.58	0.46
0.01	-0.01	0.01	-0.5	0.27	2.58	0.22	0.37	0.20
			0	0.66	2.10	0.34	0.29	0.30
			0.5	0.97	1.88	0.35	0.31	0.57
0.02	-0.02	0.01	-0.5	0.60	1.92	0.38	0.37	0.44
			0	0.82	1.51	0.32	0.33	0.40
			0.5	1.21	1.41	0.36	0.33	0.70

[a] In the first six rows, where mutation is not directional, the numbers reported are after 500 generations, and are not equilibrium values.

For all simulations, the selection function depicted in Fig. 5 was used. This is based on the observed age-specific survivorship and fecundity curves from *Drosophila*, shown in Figs. 3 and 4. The fitness of an individual with phenotype (R,D) is determined from the Euler equation. The least-squares fit to this function (with R and D in log-days) is

$$W = 0.705 + 0.107D - 0.00842D^2 - 0.102R - 0.000006R^2 + 0.0112R\,D.$$

The mutation rate for all simulations was $\mu=0.1$.

still predicts delayed senescence unless there is strong mutation correlation. (5) When directional mutation effects are added, so that most mutations cause earlier death and later reproduction (i.e., mutations have an effect opposing the selection gradient), there can be an empirical balance between directional selection and mutation. (6) Even with directional mutation, mean age of death and reproduction are positively correlated. In summary, the genetic correlations inferred in laboratory populations of *Drosophila* (Luckinbill et al. 1984; Rose and Charlesworth 1981a,b) result from a combination of selection and the pleiotropic nature of novel mutations.

Application to Metabolic Pathways

Another model that provides a mechanistic connection between a set of characters and fitness involves metabolic flux. The relationship between flux and enzyme activity has been the subject of a number of theoretical papers (Kacser and Burns 1973, 1979, 1981; Burns et al. 1985). By coupling a series of enzymes in a linear pathway, where each enzyme follows classical Michaelis-Menten kinetics, it is shown that the flux through the pathway is a rectangular function of the activity of any enzyme in the pathway. Data from the in vitro activities of enzymes in a variety of organisms show that in fact, most of the time, the activity is well up onto the "saturated" portion of the curve, indicating that large changes in enzyme activity will have relatively little affect on the flux of molecules through the pathway. Kacser and Burns (1981) argue that this is a plausible mechanism for dominance: that heterozygotes may have half the activity of an enzyme, but this may make relatively little difference to the flux through relevant pathways.

If this argument were extended to consider two enzymes imbedded in a linear pathway, one could envisage a surface describing the flux as a function of the activities of the two enzymes. Michaelis-Menten theory shows that this surface rises steeply from the origin (zero activities) to a plateau, where changes in the activity of either enzyme make little difference in flux (Fig. 6). In such a case, one might argue that any genetic variation in activities will have little affect on the organism, and hence be selectively neutral (Hartl et al. 1985). Implicit in this conclusion is that the metabolic flux can be equated with fitness. Kacser and Beeby (1984) present the argument that higher metabolic flux can confer an advantage in growth rate of primitive organisms, so that equating flux and fitness may be appropriate. Dean et al. (1986) obtained a rectangular function relating the fitness of strains of E. coli grown in lactose-limited chemostats to the activity of β-galactosidase. The experimental assessment of fitness in this case is really a measure of relative growth rate, so the results are consistent with Kacser and Beeby's (1984) model. Dykhuizen et al. (1987) present a figure based on chemostat experiments for a series of β-galactosidase and permease mutants.

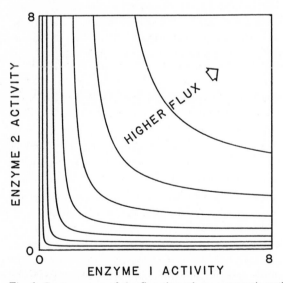

Fig. 6. Contour map of the flux through an enzymatic pathway plotted as a function of the activities of enzymes within the pathway. The function plotted is flux = 2XY/(X+Y+XY), representing equal "control coefficients" for the two enzymes. Note that as the activities of the two enzymes increase, the flux surface becomes flatter, implying that the flux is less sensitive to variations in the activity of either enzyme

The form of their results is remarkably similar to Fig. 6, having a steep ascent and a plateau in fitness for variants with high activities of both enzymes.

Once evolution has brought a population onto the plateau of the function in Fig. 6, changes in the activity of either enzyme make essentially no difference in fitness, and one would expect them to evolve independently. In spite of this, there are certain enzyme pairs that show a consistent correlation in activities. In particular, the enzymes glucose-6-phosphate dehydrogenase (G6PD) and 6-phosphogluconate dehydrogenase (6PGD) have been found to be positively correlated in activity among extracted lines of *Drosophila* (Wilton et al. 1982). The strength of this correlation was shown to be modulated by the second and third chromosome genetic backgrounds (Miyashita and Laurie-Ahlberg 1984). This is clearly at odds with the model, and suggests that either correlations are generated by pleiotropic mutation or that Fig. 6 may not be the most appropriate model relating enzyme activity and fitness. It is clear that as enzyme activity increases beyond a certain point, there will be no increase in flux, but the balance of concentrations of intermediates will be much distorted. A relation between intermediate concentration and fitness seems particularly likely for the case of G6PD and 6PGD in *Drosophila*, since double nulls are viable, but $G6PD^+$ $6PGD^-$ is lethal. Fitness may be a property of flux, but it may also depend on substrate concentrations, transient responses or costs of manufacture of enzymes (Watt 1985, 1986). Perhaps even more significant, the alteration of substrate concentrations may be very important in branched pathways, where the flux through different branches in the pathway may be quite sensitive to activities of enzymes near the branch point (Burton and Place 1986).

One strength of the approach of considering mutation-selection balance models in the context of specific selection functions is that the statement of the selection function can be fairly general, allowing an assessment of the distribution of mutations necessary to maintain the population at equilibrium. Although the success of this approach is contingent on a number of assumptions, it is rich in generating testable hypotheses about correlated characters.

A Final Constraint

No discussion of the application of quantitative genetics to evolutionary biology is complete without underscoring the experimental difficulties. The data that we would really like to have in order to test the models includes the accurate assessment of the genetic variance-covariance matrices, and this requires knowledge of the genetic relationships among a large group of individuals. Typically organisms are crossed in some well-defined design, allowing components of variance and covariance to be obtained from parents, full-sib and half-sib families. Unfortunately, this introduces the organisms to a necessarily perturbed or novel environment. Since the estimates of components of genetic covariance matrices depend on the environment (Falconer 1981; Clark 1987), the experimental results are seriously compromised.

One way around the problem of artificially induced gene by environment interaction is to rear organisms under a variety of known environmental conditions, in order to obtain a view of the norm of reaction (Mitchell-Olds and Rutledge 1986; Schlichting 1986; Via, this Vol.). But a full analysis of the evolutionary consequences of phenotypic variation may require an integration of gene-environment interaction that is not easily measurable. The real problem is that we can never know whether we have examined all of the important phenotypic characters in the requisite environmental conditions. We may stop when some reasonably clear picture emerges, but the validity of the explanations derived by this approach can always be doubted. Of course, one could argue that the understanding of any biological phenomenon (including those at the molecular level) may be compromised by failure to observe some unknown variable. It is precisely for this reason that invoking a mechanism that connects phenotypic characters together and to fitness has merit.

Summary

An important class of evolutionary constraints includes those that can be quantified as genetic correlations. The presence of genetic correlation can result in a departure of a population's trajectory from the steepest ascent up a fitness surface. Genetic correlations are themselves the product of evolution, and a number of theoretical approaches have analyzed the plausability of mutation-selection balance as an explanation for the observed levels of genetic variance. Although there is not yet a consensus on the applicability of the approximations used to solve these models, there generally appears to be more genetic variance than expected from reasonable estimates of mutation rates and strengths of selection. Although there is no such argument regarding the ability of models of mutation-selection balance to explain observed levels of genetic correlation, there really is not sufficient information to fully support either side.

The approach is extended to models where there is an explicit selection function predicted from one or more phenotypic characters. Examples include life history evolution, where fitness is a property of age-specific schedules of fecundity and mortality, and the evolution of control of enzymatic activity. In the latter case, fitness may reflect any of a number of properties of a biochemical pathway, including the rate of flux, substrate concentrations, or transient responses. The value of this approach is that it provides a means of linking an explicit selection function to known or estimatable genetic covariances, and to predict aspects of the mutational effects.

A final critical constraint that limits our understanding of phenotypic evolution is that the experimental techniques that are available (artificial selection, sib analysis, North Carolina designs, etc.) cannot be done in a natural population without perturbation. The phenotypic responses that occur when organisms are brought into the lab or planted in experimental fields may result in gene by environment interactions, resulting in a sort of uncertainty principle in components of genetic variance for phenotypic characters in natural populations. Since these estimates are needed for much of the analyses, it is dubious that a full parameterization of the models presented here will be obtained with a natural population.

Acknowledgments. I thank Drs. Ronald Burton, James Curtsinger, Anthony Dean, Russ Lande, Thomas Nagylaki, Michael Turelli, and Sara Via for stimulating discussion about many of the ideas presented here. This work is supported by NIH grant HD18379 and NSF grant BSR 8515952.

References

Antonovics J (1976) The nature of limits to natural selection. Ann M Bot Gard 63:224–247

Barker JSF (1979) Interlocus interactions: a review of the experimental evidence. Theor Popul Biol 16:323–346

Barton NH (1986) The maintenance of polygenic variation through a balance between mutation and stabilizing selection. Genet Res 48:209–216

Bulmer MG (1972) The genetic variability of polygenic characters under optimizing selection, mutation and drift. Genet Res 19:17–25

Bulmer MG (1980) The mathematical Theory of quantitative genetics, 2nd edn (1985) Clarendon, Oxford

Burns JA, Cornish-Bowden A, Groen AK, Heinrich R, Kacser H, Porteous JW, Rapoport SM, Rapoport TA, Stucki JW, Tager JM Wanders RJA, Westerhof HV (1985) Control analysis of metabolic systems. Trends Biochem Sci 10:16

Burton RS, Place AR (1986) Evolution of selective neutrality: further considerations. Genetics 114:1033–1036

Charlesworth B (1974) Selection in populations with overlapping generations. VI. Rates of change of gene frequency and population growth rate. Theor Popul Biol 6:108–133

Charlesworth B (1980) Evolution in age-structured populations. Cambridge University Press, Cambridge

Clark AG (1987) Senescence and the genetic correlation hang-up. Am Nat (in press)

Dean AM, Dykhuizen DE, Hartl DL (1986) Fitness as a function of c-galactosidase activity in *Escherichia coli*. Genet Res Camb 48:1–8

Dykhuizen DE, Dean AM, Hartl DL (1987) Metabolic flux and fitness. Genetics (in press)

Ewens WJ (1979) Mathematical population genetics. Springer, Berlin Heidelberg New York

Falconer DS (1981) Introduction to quantitative genetics, 2nd edn. Longman, London

Fisher RA (1958) The genetical theory of natural selection, 2nd edn. Dover, New York

Fleming WH (1979) Equilibrium distribution of continuous polygenic traits. SIAM J Appl Math 36:148–168

Gimelfarb A (1986) Additive variation maintained under stabilizing selection: a two-locus model of pleiotropy for two quantitative characters. Genetics 112:717–725

Gould SJ, Lewontin RC (1979) The spandrels of San Marco and the Panglossian paradigm: a critique of the adaptationist program. Proc Zool Soc Lond 205:581–598

Hartl DL, Dykhuizen DE, Dean AM (1985) Limits of adaptation: the evolution of selective neutrality. Genetics 111:655–674

Hughes DM, Clark AG (1987) Analysis of the genetic structure of life history of *Drosophila melanogaster* using recombinant extracted lines. Evolution (submitted)

Kacser H, Beeby R (1984) Evolution of catalytic proteins, or on the origin of enzyme species by means of natural selection. J Mol Evol 20:38–51

Kacser H, Burns JA (1973) The control of flux. Symp Soc Exp Biol 27:65–104

Kacser H, Burns JA (1979) Molecular democracy: who shares the controls? Biochem Sco Trans 7:1149–1160

Kacser H, Burns JA (1981) The molecular basis of dominance. Genetics 97:639–666

Karlin S (1975) General two-locus selection models: Some objectives, results and interpretations. Theor Popul Biol 7:364–398

Kimura M (1965) A stochastic model concerning the maintenance of genetic variability in quantitative characters. Proc Natl Acad Sci USA 54:731–736

Kingman JFC (1978) A simple model for the balance between selection and mutation. J Appl Probab 15:1–12

Lande R (1975) The maintenance of genetic variability by mutation in a polygenic character with linked loci. Genet Res 26:221–235

Lande R (1979) Quantitative genetic analysis of multivariate evolution, applied to brain: body size allometry. Evolution 33:402–416

Lande R (1980) The genetic covariance between characters maintained by pleiotropic mutation. Genetics 94:203–215

Lande R (1982) A quantitative genetic theory of life history evolution. Ecology 63:607–615

Lande R (1984) The genetic correlation between characters maintained by selection, linkage and inbreeding. Genet Res 44:309–320

Lande R, Arnold S (1983) The measurement of selection on correlated characters. Evolution 37:1210–1227

Latter BDH (1960) Natural selection for an intermediate optimum. Aust J Biol Sci 13:30–35

Latter BDH (1970) Selection in finite populations with multiple alleles. II. Cetripetal selection, mutation, and isoallelic variation. Genetics 66:165–186

Lewontin RC (1974) The genetic basis of evolutionary change. Columbia University Press

Lofsvold D (1986) Quantitative genetics of morphological differentiation in *Peromyscus*. I. Tests of the homogeneits of genetic covariance structure among species and subspecies. Evolution 40:559–573

Luckinbill LS, Arking R, Clare MG, Cirocco WC, Buck SA (1984) Selection for delayed senescence in *Drosophila melanogaster*. Evolution 38:996–1003

Mitchell-Olds T, Rutledge JJ (1986) Quantitative genetics in natural plant populations: a review of the theory. Am Nat 127:379–402

Miyashita N, Laurie-Ahlberg CC (1984) Genetical analysis of chromosomal interaction effects on the activities of the glucose-6-phosphate and 6-phosphogluconate dehydrogenases in *Drosophila melanogaster*. Genetics 106:655–668

Nagylaki T (1984) Selection on a quantitative character. In: Chakravarti A (ed) Human population genetics: the pittsburgh symposium. Hutchinson Ross, PA

Rose MR (1985) Life history evolution with antagonistic pleiotropy and overlapping generations. Theor Popul Biol 28:342–358

Rose MR, Charlesworth B (1981a) Genetics of life history in *Drosophila melanogaster*. I. Sib analysis of adult females. Genetics 97:173–186

Rose MR, Charlesworth B (1981b) Genetics of life history in *Drosophila melanogaster*. II. Exploratory selection experiments. Genetics 97:187–196

Schlichting CD (1986) The evolution of phenotypic plasticity in plants. Annu Rev Ecol Syst 17:667–693

Turelli M (1984) Heritable genetic variation via mutation-selection balance: Lerch's zeta meets the abdominal bristle. Theor Popul Biol 25:138–193

Turelli M (1985) Effects of pleiotropy on predictions concerning mutation-selection balance for polygenic traits. Genetics 111:165–195

Turelli M (1986) Gaussian versus non-Gaussian genetic analyses of polygenic mutation selection balance. In: Karlin S, Nevo E (eds) Evolutionary processes and theory. Academic Press, London

Watt WB (1985) Bioenergetics and evolutionary genetics: Opportunities for new synthesis. Am Nat 125:118–142

Watt WB (1986) Power and efficiency as indexes of fitness in metabolic organization. Am Nat 127:629–653

Wilton AN, Laurie-Ahlberg CC, Emigh TH, Curtsinger JW (1982) Naturally occurring enzyme activity variation in *Drosophila melanogaster*. II. Relationship among enzymes. Genetics 102:207–221

Wright S (1969) The theory of gene frequencies. In: Evolution and the genetics of population, vol 2. University of Chicago Press, Chicago

Chapter 3 Genetic Constraints on the Evolution of Phenotypic Plasticity

S. Via[1]

Introduction

In many species, dramatic phenotypic variation can be observed among individuals that are allowed to develop in different environments. Such environment-related phenotypic variation need not always indicate genetic differentiation; phenotypic variation among genetically identical individuals can result from the susceptibility to environmental influences. A change in the average phenotype expressed by a genotype in different macro-environments is generally called *phenotypic plasticity* (cf. Bradshaw 1965). Although variation due to micro-environmental effects within environments ("developmental instability", Bradshaw 1985) is considered in the models discussed here, the objective of this chapter is to discuss the evolutionary mechanisms that can produce an advantageous phenotypic response to spatial variation in the environment, that is, adaptive phenotypic plasticity.

Previous discussions of phenotypic plasticity (e.g. Gause 1947; Schmalhausen 1949; Bradshaw 1965; Schlichting 1986) have emphasized that the average level of responsiveness to the environment can evolve in a way that increases the adaptation of populations to a variable environment. The evolution of an appropriate environmental response can occur because sensitivity to the environment may differ among genotypes. Genotypes that are more susceptible to environmental influences will show a greater degree of phenotypic plasticity than do genotypes in which development is less liable to be affected by external factors. Thus, even though the proximal cues that trigger the observed phenotypic variation are strictly environmental in origin, the degree and direction of phenotypic response to the environment is usually genetically determined.

When genetic variation in susceptibility to the environment is available, an adaptive response to the environment can evolve under natural selection. Clearly, in any particular environment, only some of the possible responses to the environment will provide a selective advantage; other responses will be maladaptive interruptions of normal development. How then does "adaptive plasticity" evolve in populations?

[1] Department of Entomology and Section of Ecology and Systematics, Comstock Hall, Cornell University, Ithaca, New York 14853, USA

Genetic Constraints on Adaptive Evolution
Ed. by V. Loeschcke
© Springer-Verlag Berlin Heidelberg 1987

In this discussion of the evolutionary mechanisms of phenotypic plasticity, five questions will be considered:

1. What is phenotypic plasticity?
2. How can plasticity be measured?
3. What is the source of genetic variation in plasticity?
4. How can genetic variation in phenotypic plasticity be measured?
5. What genetic processes influence the evolution of adaptive phenotypic response to the environment?

After considering basic methods for measuring the mean and the genetic variance of phenotypic plasticity, the evolutionary dynamics of plasticity will be discussed in the context of a dynamical quantitative genetic model of evolution in a spatially variable environment. In so doing, I will attempt to smplify and expand on concepts originally presented in Via and Lande (1985). Because this genetic model is formulated in terms of genetic parameters that can be readily measured in many natural populations, it may be useful to empiricists interested in phenotypic plasticity. Several numerical examples are then provided to show how the evolution of plasticity might proceed in certain simple environmental situations.

An essential feature of the outlook presented here is that the evolution of plasticity can be studied by extending evolutionary models that concern adaptation to single environments. The major modification is that evolutionary models of a spatially patchy environment must consider not only the genetic variability available for adaptation to the phenotypic optimum in a given environmental patch, but also how different phenotypes might possibly be produced by individuals that encounter the different environments. When this is done, our models illustrate that in a coarse-grained environment (sensu Levins 1968), phenotypic plasticity (not genetic polymorphism) will result from selection toward different optima in the various patches if there is suitable genetic variability and non-zero migration among the environments. An adaptive level of phenotypic plasticity can thus be viewed as the result of stabilizing selection toward different optimal phenotypes in the range of environments encountered by a single population.

What is Plasticity?

The causes of phenotypic differences among populations have long been of interest to ecologists and evolutionists. Although there is abundant literature on this subject (reviewed for plants by Bradshaw 1965; Schlichting 1986), I will touch on only three landmark studies because they provide the conceptual framework for most of the recent work on plasticity.

1. In 1947, Gause published a summary of his experimental work on changes in body size of *Paramecium* raised in different salinities. He recognized that environmental changes in the phenotype ("modifications") are distinct from genetically based differences ("genovariation") and he was among the first to measure the selective advantage of modifications by looking at relative rates of population increase in *Paramecium* clones raised in different salinities.

2. Schmalhausen (1949) distinguished between adaptive reactions to the environment ("modifications") and non-adaptive ones ("morphoses"). He wrote that reactions to a new environment are unlikely to be adaptive initially, but that eventually, suitable variation for response to the environment would be produced by mutations which could then spread through populations by natural selection. Thus, he provided a mechanism for the evolution of adaptive plasticity. Schmalhausen (1949) also discussed the "norm of reaction" as the array or profile of phenotypes produced by a given genotype when allowed to develop in a range of environments (see Fig. 1).

3. Bradshaw's (1965) extensive review of phenotypic plasticity in plants illustrated that plasticity is a property of individual characters, not of the organism as a whole. The observation that plasticity is character-specific implies that selection may generally act on the relative values of characters expressed in different environments, not on plasticity per se.

These pivotal discussions of phenotypic plasticity yielded two working hypotheses: (1) that phenotypic plasticity is an environmentally produced change in the phenotypic expression of a genotype that may or may not be adaptive, and (2) that plasticity can evolve under natural selection in a variable environment by providing a selective advantage for organisms that produce an appropriate environmental response.

Nearly all characters can exhibit some type of phenotypic plasticity. In an attempt to capture the diversity of environmental modifications of the phenotype, I will list several major categories of plasticity, with just a few examples to illustrate each one. (1) *Morphological plasticity* includes the classic example of leaf-shape

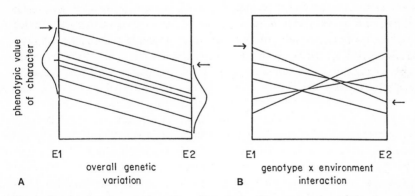

Fig. 1. Graphical representation of reaction norms. The mean value of the character state expressed by a given genotype in Environment 1 (*E1*) is plotted on the left axis, the mean phenotype expressed in Environment 2 (*E2*) is plotted on the right axis, and the genotypic means in the two environments are connected by a line. The slopes of the lines thus denote the phenotypic response of each genotype to a change in environment. Within environments, variation among genotypes is expected to be normally distributed. **A** All genotypes respond similarly to a change in the environment; the reaction norms are parallel and no changes in genotypic rank occur. **B** The variation among genotypes within environments is the same as in **A**, but genotypes vary in their response to the environment. This genetic variation in reaction norms is called "genotype-environment interaction". *Arrows* mark certain genotypes to show that changes in rank occur in **B** but not in **A**

modification in aquatic plants (e.g. Bradshaw 1965), induced defenses in rotifers, bryozoa, and ciliates (Gilbert 1966; Harvell 1984; Kuhlmann and Heckmann 1985), and changes in the musculature of the head with different diet hardness in grasshoppers (Bernays 1986). (2) *Physiological plasticity* is also common, including differences in photosynthetic rates of plants in sun and shade environments (Clough et al. 1979) and seasonal changes in water balance (Teeri 1978). In animals, acclimation to changes in elevation or temperature is widespread (Prosser 1973). (3) Learning is the premier example of *behavioral plasticity,* and animals of all types exhibit some degree of conditioning or learning in their behaviors. (4) *Biochemical plasticity,* such as changes in the activity or conformation of proteins with a change in the intracellular environment (i.e., pH or temperature), illustrates plasticity at the subcellular level (Hochachka and Somero 1984).

Clearly, plasticity at some levels may permit constancy at others. For example, either behavioral thermoregulation or biochemical plasticity can permit physiological homeostasis. There is no single level at which plasticity should be expected (see Orzack 1985, for a discussion of plasticity ahd homeostasis).

Sometimes plastic changes in entire suites of characters can occur, as, for example, the coordinated changes in size, coloration, physiology, and flight behavior that distinguish gregarious and solitary phases of locusts (Nijhout and Wheeler 1982). The striking differences between locust phases can be traced to population density during development. Phenotypic responses to the environment may either involve permanent modifications if environmental sensitivity is restricted to a limited period during development (e.g. Wheeler and Nijhout 1981), or may be reversible, as in physiological acclimation. Finally, plasticity may involve responses to either biotic or abiotic features of the environment.

Although visible changes in the phenotype must in some way be linked to changes in gene expression or the differential sensitivity of tissues to certain gene products in various environments, precise molecular and developmental mechanisms for most cases of phenotypic plasticity are currently unknown. There are, however, some examples of changes in the expression of genes in different environments, such as the production of heat-shock proteins under stress conditions (Atkinson and Walden 1985). Although at present, studies of biochemistry or cell biology provide only a framework of possibility within which to speculate, basic genetic and developmental research may eventually provide a more mechanistic understanding of how changes in the external environment are transduced into plastic modification of the visible phenotype.

Measuring Phenotypic Plasticity

Despite agreement that the extent of phenotypic plasticity can evolve in populations, a metric for plasticity that permits a precise evolutionary interpretation has remained elusive. Many different methods for measurement have been suggested, each reflecting a slightly different view of plasticity. For example, (1) Khan et al. (1976) measured the responses of flax and linseed to changes in plant density by computing the heritability of the ratio of phenotypes when sib groups were raised in different densities. (2) Jain (1978) measured plasticity in the plant *Bromus mollis* as the increase in within-family variance of sibling groups grown in a range of different environments. (3) In a study comparing the plasticity of two species of *Phlox,* Schlichting and Levin (1984) used two measures of plasticity. The *amount* of plasticity for each species was measured as the coefficient of variation (CV) of the treatment means when 20 plants from each species were grown in each of 6 different treatment environments (CV = standard deviation of treatment mean/mean of treatments). The *pattern* of plasticity was calculated as the correlation between the treatment means of the two species. Note that in this experiment, the plants raised in each treatment were independent samples from the same populations, *not* the same collection of genotypes. Therefore, no formal genetic estimates can be derived from such a design. (4) Using a variance-components approach, Scheiner and Goodnight (1984) estimated the "plasticity variance" as the sum of the genotype-environment interaction, V_{GxE} (the variance among genotypes in phenotypic plasticity), and the environmental variance, V_E (the average phenotypic plasticity). This method does require that the same genotypes (or siblings from the same families) be tested in each of the environments. However, the evolutionary inferences that can be made from adding the mean and the variance of phenotypic plasticity are somewhat unclear.

Moreover, although these methods for measuring plasticity provide useful descriptions of the nature of phenotypic variation within species, they are not in a form that can be readily utilized in a mathematical description of the process of evolution. Therefore, they are not as useful as the method of genetic correlations discussed below, either for the study of how evolutionary processes might have produced current levels of plasticity or for the estimation of the potential for future evolution of a different level of plasticity.

In order to understand how adaptive evolution might occur in any trait, it is useful to have a precise theoretical context within which the effects of particular measurable genetic and ecological factors can be studied. One of the goals of this chapter is to discuss a quantitative genetic theory for how the norm of reaction might evolve. To formulate this theory (discussed in detail in Via and Lande 1985), it is useful to measure plasticity as the difference in the mean phenotypic values that are expressed in several environments by clonal replicates (or siblings). This method essentially provides an experimental measure of the norm of reaction for each genotype (or family). The evolution of plasticity can then be studied by considering how the norm of reaction can evolve.

Given that each genotype has a norm of reaction (that is, produces some possibly different phenotype in each of several environments), then an population is characterized by (1) a mean norm of reaction (the average response to the environment) and (2) some genetic variation in the norms of reaction. With this logic, it can be hypothesized that the norm of reaction evolves like any other quantiative trait, at a rate and in a direction that depends on both the shape of the current mean reaction norm relative to some "optimum norm" and on the magnitude of genetic variation in reaction norms.

The role of genetic variability in reaction norms in the evolution of phenotypic plasticity can be seen with a simple example. Consider a population in which individuals may encounter either of two environments (E1 or E2). In Fig. 2, the

NORM OF REACTION

Profile of phenotypes produced in different environments

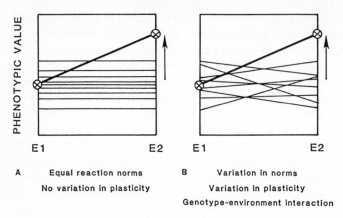

<table>
<tr><td>A</td><td>Equal reaction norms</td><td>B</td><td>Variation in norms</td></tr>
<tr><td></td><td>No variation in plasticity</td><td></td><td>Variation in plasticity</td></tr>
</table>

Genotype-environment interaction

Fig. 2. Two hypothetical populations with different amounts of genetic variation in phenotypic plasticity. Genetic variation within environments is the same for all cases. *Thin lines* are reaction norms for individual genotypes. The new optimal norm of reaction is drawn as a *bold line,* with new phenotypic optima in each environment denoted as ⊗. Thus, some new force of selection favors a larger mean phenotype in *E2* than in *E1*. In **A**, where there is no variation among genotypes in the norms of reaction, the new norm can never evolve. In **B**, genetic variation in reaction norms is present, so the phenotypes expressed in each environment can evolve somewhat independently and the population can eventually evolve to a new level of phenotypic plasticity. If selection is weak the genetic variance within environments is expected to remain roughly constant (Via and Lande 1987), while the mean phenotype evolves to the new value in *E2*

thin lines connect the average phenotypic values produced by various genotypes when clonal replicates or siblings are allowed to develop in the different environments. The slope of each line is a visual representation of the plasticity (change in phenotype across environments) of each genotype; these lines are the norms of reaction. In Fig. 2A and B, the mean phenotype produced by the hypothetical population is the same in both the environments, that is, the *mean* reaction norm for the population has a zero slope.

Imagine now that this hypothetical example concerns a plant population in which the environment has changed, making it advantageous to be taller in one environment than in another. One possible environmental change with this effect might be a release from the selection for a prostrate phenotype that can be caused by trampling or grazing (Briggs and Walter 1984, Chap. 13). This change in the environment leads to a new optimum norm of reaction in which the phenotypic optimum is larger in one environment (E2 in Fig. 2) than in the other (E1 in Fig. 2). This new optimal norm is denoted by the heavy lines in Fig. 2. In other words, a taller phenotype is favored among seeds that disperse into E2 than among those that disperse into E1. Evolution toward this new norm requires an increase in the average plant height in E2, while holding height constant in E1. We are interested in how genetic variation in plasticity can affect this evolutionary process.

For a population with the reaction norms shown in Fig. 2A, attainment of the new reaction norm is impossible because genotypes with a higher than average value of the phenotype in Environment 2 will also show an increased height in Environment 1, putting them at a disadvantage there. In contrast, in the population illustrated in Fig. 2B, variation among genotypes in the response to the environment will allow the eventual evolution of a new reaction norm.

When there is genetic variation in phenotypic plasticity (genetic variation in reaction norms), the mean phenotypes expressed in the two environments can evolve semi-independently. In contrast, when all genotypes respond to the environment in exactly the same way, no evolutionary change in the relative mean phenotype expressed in two environments is possible. In other words, genetic variation in reaction norms is necessary for the evolution of a new norm of reaction, and consequently, for a change in the extent of phenotypic plasticity expressed by a population.

How does Genetic Variation in Reaction Norms Arise?

We know that genes can be expressed differently in different environments; temperature-sensitive alleles are an obvious example that allelic variants can have very different environmental sensitivities. Schmalhausen (1949) recognized that since mutations may differ from wild type in not only the average phenotype, but also in the responsiveness to the environment, mutation serves as an important source of variation in the norm of reaction. He wrote that "mutation is a *change in (the) reaction norm*" (p. 10, italics in original), and that

"The organization of plants and animals, which is based upon mutation, becomes adaptive only in the process of evolution. Similarly, all reactions of living matter, which are also based upon mutations, become adaptive only in the course of historic development in certain environments. Everything new, including a capacity for adaptive modification, is created in the process of natural selection of mutations together with their characteristic norms of reaction" (p. 8).

As mutations occur in populations, they therefore generate variation not only in the values of character states expressed in particular environments (e.g. Lande 1980; Turelli 1984), but also in the norms of reaction. In fact, the observed covariance between character states can be shown to be largely a function of the covariance of mutational effects on the phenotype (Via and Lande 1987). As long as mutation generates variance in the norms of reaction, new levels of phenotypic plasticity can evolve.

Measuring Genetic Variation in Phenotypic Plasticity

Figures 1B and 2B illustrate differences among genotypes in their responses to the environment. Such differences are often quantified by measuring genotypes (or family members) in several environments and then performing a two-way ANOVA in which "genotype" and "environment" are the factors. Variation among genotypes in phenotypic plasticity is then estimated from the observed genotype-environment interaction (for review of statistical methods, including the "regression approach", see Freeman 1973). Note that the g-e interaction has a precise statistical definition as the variation among genotypes in response to the environment; it is *not* just the "collaboration" of genotype and environment in the production of the phenotype.

Although the g-e interaction may be the traditional way to study variability in environmental response, it is not necessarily the most informative method. As I will describe below, g-e interaction does not provide a precise view of the genetic variation that is available for evolutionary change in the relative phenotypes expressed in two environments. An alternative statistic, the genetic correlation between the character states expressed in different environments, is a more useful metric because it precisely quantifies genetic variation in plasticity in a way that can be readily interpreted in genetic models of evolution.

Falconer (1952) first recognized that one character expressed in two environments can be thought of as two *character states,* each of which is expressed in only one of the environments. For example, in the situation discussed above, plant height could be considered to be two character states, "plant height in E1" and "plant height in E2". Each character state has a measurable genetic variation within the environment in which it is expressed. Moreover, the states expressed in the two environments may share some common genetic basis that would produce a "genetic correlation" (denoted as r_G) between them.

In general, the additive genetic correlation between two characters measures the extent to which they share a common genetic basis (Falconer 1981). In the case of two character states, this means that they will be highly genetically correlated if the development of the character states expressed in each of two environments involves expression of most of the same genes in the same way. A high additive genetic correlation between character states corresponds to a low degree of genetic variation in phenotypic plasticity. As the additive genetic correlation between two character states approaches ±1, their potential for independent evolution goes to zero; a high correlation means that many gene frequency changes in one character will also affect the other proportionately. The greatest possibility for a change in the *relative* values of two character states, that is, for a change in phenotypic plasticity, exists when the genetic correlation between the states is zero ($r_G = 0$).

The genetic correlation between two character states determines the magnitude of the genotype-environment interaction that would be estimated in an analysis of variance (Robertson 1959; Yamada 1962; Fernando et al. 1984; Via 1984b). Assuming for simplicity that the variance among genotypes (V_G) is the same within each of the two environments, and that the data set is balanced, the g-e interaction (V_{GxE}) can be written in terms of the genetic correlation between character states by modifying Yamada's (1962) formulation as

$$V_{GxE} = V_G(1 - r_G). \tag{1}$$

From Eq. 1 we can see that the interaction variance increases monotonically as the genetic correlation between the character states expressed in the two environments declines from +1 to −1, even though the greatest opportunity for an evolutionary change in phenotypic plasticity exists when $r_G = 0$. This is why the g-e interactions does not provide a precise picture of the potential for evolution of phenotypic plasticity.

Since an average value of a given character state can be measured for each family in each environment, the pairwise genetic correlations between the character states expressed in different environments can be calculated from the product-moment correlation of family means (see Via 1984b, for discussion of other methods). This metric for genetic variation in plasticity is very useful theoretically. Just as the evolution of a suite of traits in a single environment depends on their genetic correlations (Lande 1979), so the evolution of a set of character states expressed in different environments depends critically upon the pattern of genetic correlations between states (Via and Lande 1985).

For comparative purposes, Fig. 3 shows the same data set plotted using both g-e interaction plots and scatter diagrams that depict the genetic correlation between two character states. The usual genotype-environment interaction plots are drawn in the left-hand panels; as before, the phenotypic value of each genotype (or family) in the two environments is connected by a thin line. Going from the top to the bottom of Fig. 3, the plots depict populations with increasing magnitudes of genotype-environment interaction variance.

When the character expressed in each environment is considered to be a separate character state, then the same data graphed in the left-hand panels of Fig. 3 can be replotted as correlation plots. These are shown in the right-hand panels (Fig. 3B, D, F, H). The value of the character state expressed in Environment 1 is denoted as z_1; the phenotypic value expressed in Environment 2 is z_2. Thus, instead of a line (reaction norm) for each genotype, the scatter plots show the average value of the phenotype expressed by a given genotype in each environment as a point in z_1-z_2 space. From the correlation of these clonal (or family) means, the genetic correlation between the character states expressed in the two environments can be calculated (Via 1984b). When the relative phenotype expressed in the two environments is graphed on a scatter plot as in Fig. 3B, D, F, H, phenotypic plasticity is the difference in the average phenotype expressed in the two environments (in this case, there is initially no plasticity). Genetic variation on phenotypic plasticity is measured as the covariance of character states expressed in the two environments.

Comparing the right and left panels of Fig. 3, it can be seen that when reaction norms in a population are parallel (Fig. 3A), the genetic correlation among the traits is +1 (Fig. 3B). This suggests that the same alleles or sets of alleles affect the character states in the two environments proportionally, so that genotypes maintain their relative ranking in the two environments. As the norms of reaction within a population show greater variation, that is, greater g-e interaction (Fig. 3C, E), the genetic correlation between character states decreases (Fig. 3D, F). When $r_G = 0$, the genetic variance available for a change in the relative value of the phenotype expressed in the two environments is at a maximum (Fig. 3F). However, as the genetic correlation between the character states decreases to -1 (Fig. 3H), the g-e interaction variance continues to increase (Fig. 3G), even though the genetic variation available for a change in the relative value of the phenotypes expressed in the two environments declines to zero. Thus, the genetic correlation between

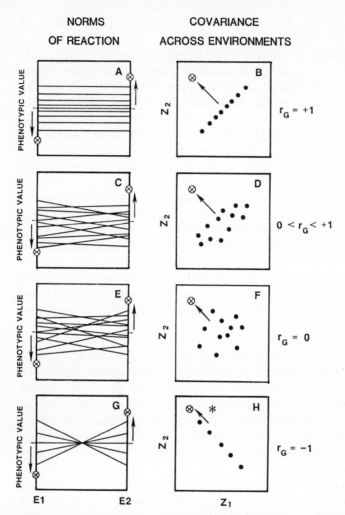

Fig. 3. Four levels of genetic variation in phenotypic plasticity plotted in two ways. **A, C, E, G:** Genotype-environment interaction plots. **B, D, F, H:** Scatter plots from which the genetic correlations between character states can be estimated. Each *line* or *point* corresponds to the mean phenotype that a given genotype expresses in each of two environments. Axes for the *left-hand panels* are as in Fig. 1. In the *right-hand panels*, z_1 is the character state that a given genotype expresses in Environment 1, z_2 is the character state expressed by that genotype in Environment 2, and r_G is the value of the additive genetic correlation between character states expressed in *E1* and *E2*. In all plots, the mean phenotype expressed in each environment is the same, that is, the average level of phenotypic plasticity is initially zero. ⊗ indicates a new optimal level of phenotypic plasticity toward which the population is being selected, with stabilizing selection on the two character states in the direction of the *arrows*. Genetic variation for a change in the current level of phenotypic plasticity is zero in **A, B, G,** and **H** ($r_G = \pm 1$), intermediate in **C** and **D** ($0 < r_G < 1$), and high in **E** and **F** ($r_G = 0$), although the magnitude of the g-e interaction variance increases monotonically from top to bottom. New optima ⊗ or ∗ will never be attained in **B** or **H**, respectively (see text)

character states expressed in different environments accurately reflects the available genetic variation for the evolution of plasticity, while the g-e interaction does not.

The evolutionary effects of the correlation across environments can be graphically illustrated with the plots in Fig. 3. Consider aquatic plants that exhibit heterophylly (broad aerial leaves, narrow aquatic leaves). If a plant population originally produces leaves of equal width in water and air, how can a difference in the leaf width expressed in aerial and aquatic environments evolve? In Fig. 3, selection to decrease the value of the phenotype (leaf width) in E1 and to increase it in E2 is indicated by arrows toward the new optimum in each environment, denoted as ⊕.

As seen in Fig. 2, when there is no genetic variation in reaction norms, most new optima will usually not be attained. Figure 3 shows why this is true. When the genetic correlation between character states is ±1, all the points fall on a line; only a new optimum that is *exactly* on this line can be attained (e.g., Fig. 3H). There is no genetic variation for any other combination of phenotypes in the two environments. Therefore, the population in Fig. 3B could not respond to selection to move the average phenotype in the direction of the arrow. Although the population in Fig. 3H could attain the optimum in this particular example, because it happens to be on the line, it must be noted that if the new optimum corresponded to any pair of phenotypes that were even slightly off that line, such as *, it could never be attained, because no genetic variation is available for any phenotypic combinations except those that are exactly on the line.

In contrast, when the genetic correlation between states is less than ±1 (Fig. 3D, F), there is some genetic variation available for a response to selection to move the joint mean phenotype in any direction in phenotype space. Moreover, all else being equal, we would expect the new optimum to be attained more rapidly in a population like Fig. 3F than in one like Fig. 3D because the lower genetic correlation permits a greater degree of independent evolution of the two character states, that is, there is more genetic variability in the direction of the new optimum.

In general, the reason that genetic correlations between characters can be constraining is that less genetic variation is available for evolution in some directions (in the direction of the narrow part of the ellipse of points in Fig. 3B, D, F, H) than in other directions (along the long part of the ellipse).

A Quantitative Genetic Model for the Evolution of Phenotypic Plasticity

The formulation of a mathematical model of evolution can be a useful way to determine how various ecological and genetic factors can influence the course of evolution. In the previous section, we saw how a difference in the average phenotype expressed in different environments and the genetic covariance between the character states measures the mean and genetic variance of plasticity in natural populations, respectively. Now we will show how these parameters can be used in a mathematical model that describes the dynamics of evolution of plasticity. This model will show that some ecological or genetic factors can slow or prevent the evolution of an appropriate response to the environment (for a detailed description of these models, see Via and Lande 1985). The genetic correlation between character states expressed in different environments is a particularly useful metric for genetic variation in phenotypic plasticity because it allows the dynamics of the evolution of plasticity to be accessed with the same kind of mathematics that was originally developed to describe evolution of a correlated suite of characters within a single environment (e.g. Lande 1979).

The models discussed here are formulated for a simple population structure (Fig. 4). We assume a large randomly mating population in which some individuals experience each of two environments, for example, an insect population that feeds on two co-occurring host plant species, but does not mate on the host. Within each environment, stabilizing selection acts around some optimum phenotype that can differ for the two environments (Fig. 5). Most ecologically important traits have intermediate optimum phenotypes; this model thus applies to many morphological,

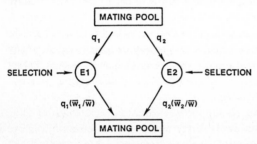

Fig. 4. Selection scenario. Random mating occurs, followed by movement of zygotes into the environments; there is no habitat selection. A fraction of the population, q_1, goes to the first environment and the rest, $q_2 = (1-q_1)$, goes to the second environment. Stabilizing selection to a possibly different optimum value occurs within each environment, after which the portion of the population that experienced the ith environment has a mean fitness of $\overline{W}_i/\overline{W}$. Then, individuals re-enter the random mating pool in proportion to the environmental frequency and the mean fitness in each environment

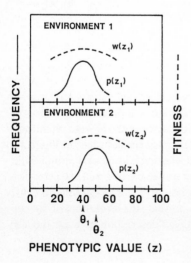

Fig. 5. Stabilizing selection toward different phenotypic optima in two environments (optimum in E1 is 40 units, optimum in E2 is 50 units). In the ith environment, the *dashed line* is the stabilizing selection function, $w(z_i)$; fitness declines for phenotypes that deviate from the optimum value, θ_i. The *solid line* is the phenotype distribution, $p(z_i)$. The force of selection is proportional to the width or curvature of the fitness function (the flatter the fitness function, the weaker the selection) and the deviation of the average phenotype (z) from the optimum (θ)

physiological, and behavioral characters. However, this model does *not* apply to major components of fitness that can be presumed to be under a continual force of directional selection. In general, the characters of interest to ecologists are polygenic and can usually be measured on a scale of measurement that yields a normal distribution, with the mean approximately independent of the variance (Falconer 1981).

Each generation after mating, zygotes are presumed to be distributed across the environments in proportion to the fraction of the total environment composed of each habitat type; there is no habitat choice. Stabilizing selection acts on the offspring within each environment. Finally, any individual experiences only one of the environments, that is, the environment is "coarse-grained" (sensu Levins 1968).

Because only part of the population experiences each environment, the evolution of the overall phenotype is a weighted sum of the responses to selection in each environment. The weighting factor has two parts: (1) the proportion of the population that experiences selection in each habitat (q_i), and (2) the relative mean fitness of individuals that experience that environment (e.g., $\overline{W}_i/\overline{W}$ for the ith environment, where $\overline{W} = \Sigma\, q_i\overline{W}_i$). After selection, each environment contributes individuals to a randomly mating pool in proportion to $q_i(\overline{W}_i/\overline{W})$. The major question that we will address using these models is how the evolution of a different mean phenotype in the two environments can proceed when part of the population experiences each environment every generation and yet is mixed with the other portion of the population in a random mating pool.

The essential logic of the models can be understood by extending the usual expression for the response to selection in a single environment to apply to responses of character states to selection for a (possibly) different mean phenotype in each of several environments. The standard univariate response to selection equals the product of the genetic variability and the intensity of selection on the character, that is, $R = h^2 s$, where R is the response to selection, h^2 is the heritability of the trait of interest, and s is the difference in the average phenotype before and after selection (Falconer 1981).

This basic equation for the response to selection was generalized for multiple characters by Lande (1979), as $\Delta\overline{z} = G\, P^{-1}\, s = G\nabla\ln\overline{W}$, where $\Delta\overline{z}$ is the vector of changes in the characters of interest in one generation (the responses), G is a matrix of genetic variances and covariances among the traits, P is the matrix of phenotypic variances and covariances (GP^{-1} is the multivariate analog of heritability), s is a vector of selection differentials, and $\nabla\ln\overline{W}$ is the vector of forces of selection acting directly on each character [$\nabla = (\partial/\partial\overline{z}_1, \partial/\partial\overline{z}_2)$, meaning that $\nabla\ln\overline{W}$ is the vector of changes in mean fitness due to changes in each character when all others are held constant]. Thus, Lande's equation is the multivariate extension of the univariate formula $R = h^2 s$, assuming that character states are distributed as a multivariate normal (meaning that a bivariate scatter plot for two character states would look roughly like Fig. 3B, C, or F), and that all genetic variation is additive.

Via and Lande (1985) further extended this modeling approach to describe the evolution of a character expressed in two environments as

$$
\begin{pmatrix} \Delta\bar{z}_1 \\ \Delta\bar{z}_2 \end{pmatrix} = \begin{pmatrix} G_{11} & G_{12} \\ G_{12} & G_{22} \end{pmatrix} \begin{pmatrix} q_1(\bar{W}_1/\bar{W})\,\nabla_1\ln\bar{W}_1 \\ q_2(\bar{W}_2/\bar{W})\,\nabla_2\ln\bar{W}_2 \end{pmatrix}, \tag{2}
$$

where $\Delta\bar{z}_i$ is the change in the character state expressed in the ith environment in one generation of selection, $q_i(\bar{W}_i/\bar{W})\nabla_i\ln\bar{W}_i$ is the force of selection on each character state with a weighting as described above (micro-environmental variance within environments is incorporated in this term, see Via and Lande 1985), G_{ii} is the genetic variance of the character state that is expressed in the ith environment, and G_{ij} is the genetic covariance between the character states. The genetic correlation between the states is thus $G_{ij}/\sqrt{G_{ii}G_{jj}}$.

Note that this method can be used to study more characters or environments by expanding the dimensions of the matrix and vectors. When a group of characters is examined in each environment, the overall **G** matrix will be partitioned, with blocks on the diagonal that contain the covariances within environments and blocks off the diagonal that contain the covariances of the character states in different environments. Because the covariances for each environment are written in a separate block, this method takes into account the possibility that genetic correlations may differ within different environments (e.g. Via 1984b; Schlichting 1986).

When the matrix equation in Fig. (2) is multiplied out, the evolutionary changes in the character states are

$$
\Delta\bar{z}_1 = q_1(\bar{W}_1/\bar{W})G_{11}\,\nabla_1\ln\bar{W}_1 \;+\; q_2(\bar{W}_2/\bar{W})G_{12}\,\nabla_2\ln\bar{W}_2; \tag{3a}
$$

$$
\Delta\bar{z}_2 = q_2(\bar{W}_2/\bar{W})G_{22}\,\nabla_2\ln\bar{W}_2 \;+\; q_1(\bar{W}_1/\bar{W})G_{12}\,\nabla_1\ln\bar{W}_1. \tag{3b}
$$

$$\underbrace{\hspace{4cm}}_{\text{direct response}} \qquad \underbrace{\hspace{4cm}}_{\text{correlated response}}$$

The dynamics of evolution in each character state thus has two components. The *direct response* is the response to selection on the character state in the environment in which it is expressed. Equations (3a) and (3b) show that the magnitude of the direct response depends on the frequency of the environment in which the character state of interest is expressed (q_i), the relative mean fitness of individuals in that environment ($\bar{W}_i/\bar{W}$), the genetic variance of the character state of interest (G_{ii}), and the force of stabilizing selection within that environment ($\nabla_i\ln\bar{W}_i$). The *correlated response* to selection for either character state is the amount of change in that state due to selection on the correlated character state expressed in the part of the population that was selected in the other environment. For example, even though z_1 is not expressed in Environment 2, individuals in that environment carry genes for z_1. To the extent that z_1 and z_2 are genetically correlated (indicated by the sign and magnitude of G_{12}), gene frequencies at loci affecting z_1 can be changed by selection acting on z_2.

Thus, if one is interested in the evolution of a character expressed in a particular environment, it is important to consider not only the response to selection in that environment, but also the possibility of correlated responses to selection that could occur in other environments experienced by the population. Any factors that increase the magnitude of a correlated response to selection that is opposite in sign to the direct response will constrain the evolution of phenotypic plasticity by slowing the rate at which populations attain the optimum phenotype in one or both environments. Equations (3a) and (3b) suggest that the following genetic and ecological factors could constrain the evolution of phenotypic plasticity:

1. *The Relative Magnitudes of the Genetic Variances.* If, for any reason, the genetic variance of the character state expressed in one of the environments is smaller than that of the other character state, evolution of the state with the smaller variance may be dominated by a correlated response to selection on the other state.

2. *The Sign of the Genetic Correlation Between Character States Relative to the Direction of Selection.* Given equal environmental frequencies and genetic variances, the response to selection will be slowed if (i) both character states are selected to increase and the genetic correlation is negative, or (ii) one character state is selected to increase, the other state is selected to decrease, and the genetic correlation between the states is positive. In either of these two cases, Eq. (3) shows that the correlated response will be opposite in sign to the direct response, thus reducing the magnitude of the overall response to selection.

3. *The Relative Frequencies of the Environments.* Evolution will proceed very slowly in a rare or marginal environment because few individuals experience rare environments (q_i small) and relative mean fitnesses may be low in marginal ones ($\overline{W}_i/\overline{W}$ small). Selection is very ineffective in such environments because few individuals are contributed to the mating pool. All else being equal, in a rare or marginal environment the direct response to the environment will be smaller than the correlated response because the weighting factor $[q_i(\overline{W}_i/\overline{W})]$ will tend to be small in such a situation.

4. *The Strength of Selection.* Under stabilizing selection, the strength of selection is influenced both by the distance of the mean phenotype from the optimum phenotype and by the width of the stabilizing selection function (see Fig. 5 and Via and Lande 1985). If the forces of selection in the two environments differ ($\nabla_1 \ln\overline{W}_1/\nabla_2 \ln\overline{W}_2$) for either of these reasons, then evolution will occur more rapidly where selection is stronger (all else being equal), possibly swamping the direct response to selection in the other environment.

Numerical Examples

In order to illustrate the behavior of the models and to visualize how changes in the parameters can affect the evolution of plasticity, it is useful to look at some numerical examples of evolutionary trajectories for populations in certain well-defined circumstances. These examples show how correlated responses to selection can cause maladaptation in the character state that is expressed in the environment in which evolution proceeds most slowly. First, I will describe two examples of how the genetic correlation among character states can affect the evolution of an adaptive level of phenotypic plasticity for populations selected in two environments. Then I will expand on these examples to probe how selection in multiple environments can cause the constraining effects of genetic correlations to be exacerbated.

In order to focus on the effects of the genetic correlations between character states on the evolution of plasticity, we establish some symmetries for the rest of the parameters of the models. Note, however, that the symmetrics used here are for purposes of illustration; they are not necessary assumptions of the models. Moreover, *any* changes in the parameters that cause unequal rates of evolution in the two environments could lead to constraints similar to those generated in these examples.

For the examples, we first assume that the genetic and phenotypic variances of the character states expressed in each environment are equal ($G_{11} = G_{22} = G$ and $P_{11} = P_{22} = P$, with $P - G = E$, the micro-environmental variance within environments). We also assume that the width of the stabilizing selection function within each environment is the same, that is, that individuals with phenotypes that deviate from the optimum suffer an equal fitness penalty in each environment. Finally, we assume that the matrix of genetic variances and covariances remains roughly constant during the evolution of the mean phenotype after a perturbation. Via and Lande (1987) show that this assumption is generally valid unless the population of interest is perturbed very far from the joint optimum or has a history of frequent disturbance.

It is important to note that in our models, the optimum phenotypes in the two environments can be any pair of values. When the values of the optimum phenotypes in the two environments are different, there will be selection for phenotypic plasticity. Even if the optima in the two environments are the same, however, the evolution of the joint phenotype will still be described by the same genetic model. Selection toward equal optima in the two environments is just a special case of the general model of evolution of quantitative traits in a heterogeneous environment.

Example 1: Two Environments, One is Rare

Imagine an insect population that feeds on several host plant species, one of which is rare in a given locality. The expected evolutionary trajectories for a population that experiences two environments of unequal frequency ($q_1 = 0.7$, $q_2 = 0.3$) are shown in Fig. 6A. In all of the following examples, the value of the mean phenotype expressed in Environment 1 is denoted as $\bar{z}_1$, and the value of the mean phenotype expressed in Environment 2 is written as $\bar{z}_2$. In the case of the hypothetical insect

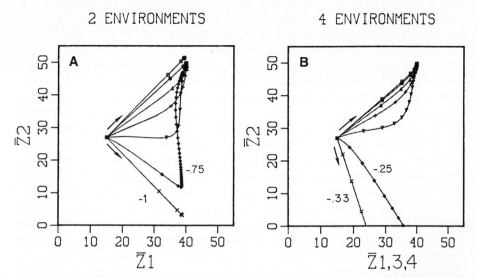

Fig. 6. Effects of the genetic correlation between character states expressed in different environments on the trajectory of the mean phenotype when environments are represented in unequal frequencies (q_1 = 0.7). $\bar{Z}_1$ is the mean phenotypic value of the character state expressed in E1, and $\bar{z}_2$ is the mean phenotype of the character state expressed in E2. Each curve is the trajectory of a population with a different value of the genetic correlation between character states as it evolves from the initial point to the optimum: +1 (□); +0.75 (○); +0.375 (△); 0 (+); −0.375 (✕); −0.75 (◆); −1 (▽). Symbols are marked every 50 generations. Phenotypic optima are 40 in E1 and 50 in E2, and evolution occurs in the direction of the *arrows*. Initially, the joint mean phenotype is at $\bar{z}_1$ = 15 and $\bar{z}_2$ = 27. **A** Two environments. **B** Four environments, one is rare. Environments 1, 3 and 4 are present at equal frequency (q_1 = q_3 = q_4 = 0.28). See text for details

population that we are considering in this example, the phenotypic trait, z_i, might correspond to a character such as the average body size attained by the segment of the population that experiences the ith environment. For this example, the optimum joint phenotype is arbitrarily set at (40, 50), meaning that it is advantageous to have a slightly larger phenotypic value in Environment 2 than in Environment 1. This corresponds to a situation in which the best norm of reaction would be to show a slight degree of phenotypic plasticity toward larger size in E2.

Now imagine that the population is either perturbed from the joint phenotypic optimum or that the optimum changes such that the mean phenotype expressed in both environments is too small. Figure 6A illustrates how populations characterized by particular values of the genetic correlation between the character states expressed in the two environments will evolve back to the optimum level of phenotypic plasticity (i.e., toward the point 40, 50).

As expected from Eqs. (3a) and (3b), evolution occurs more rapidly in the common environment than in the rare one. With no habitat selection, few individuals experience the rare environment and so selection is less effective there. Thus, the evolutionary dynamics of the character state expressed in the rare environment are dominated by the correlated response to selection in the common environment. Figure 6A illustrates that in this situation, certain values of the genetic correlation between the character states can have a large impact on the evolutionary trajectories. If the genetic correlation between the states is positive, then the correlated response of the character in the rare environment will move it toward the optimum just as the direct response would have if q were larger. However, if the genetic correlation between the states is negative, then adaptation of the character in the common environment occurs at the expense of the state expressed in the rare environment. Figure 6A shows that when the genetic correlation is opposite in sign to the direction of selection, the character state expressed in the rare environment can actually evolve away from its optimum in a maladaptive fashion. This process of maladaptation will halt only when the force of selection to move the character state in the rare environment back to its optimum increases to the point at which it counterbalances the diminishing force of selection on the state expressed in the common environment. Note that a mirror image of Fig. 6A would be observed if selection were acting to move the characters expressed in the two environments in opposite directions. Then, *positive* genetic correlations would cause maladaptation in the rare environment.

Example 2: Two Environments, One is Novel

In the next example, we will use the models in Eqs. (2) and (3) to illustrate the course of evolution of phenotypic plasticity when a population is exposed to a novel environment (one to which it is not initially well adapted). The problem of evolutionary interest here is how a change in the phenotype expressed in the new environment can be effected while retaining the well-adapted penotype in the "traditional" environment. Thus, a new level of phenotypic plasticity is required. Consider an insect population that is expanding its range and encounters a new host plant in the same locality as another host plant species to which it is already well adapted (i.e., on which it produces a phenotype that is close to the optimum phenotype). This population is selected for phenotypic changes that increase adaptation to the new host species without disrupting the phenotype expressed on the old host. Although this situation is formally equivalent to a shift in the phenotypic optimum (perhaps by introduction of a new predator or competitor in one of two "traditional" environments), we will pursue the "novel environment" case for the purposes of illustration.

In this hypothetical insect population, the dynamics described by (3) are dominated by the fact that selection is stronger on the new host because the mean phenotype in the population is further from the phenotypic optimum for that environment. If we assume that all else is equal, the greater force of selection in the new environment causes evolution to be more rapid there. If the character states expressed on the two hosts are genetically correlated, then the character state expressed on the old host can be changed due to a correlated response to selection on the character state expressed on the new host, possibly evolving away from its individual optimum (e.g., $r_G = -0.75$ in Fig. 7A). Maladaptive evolution can occur for high genetic correlations of either sign (Fig. 7A). When the genetic correlation is positive, the correlated response to selection causes the phenotype in the old

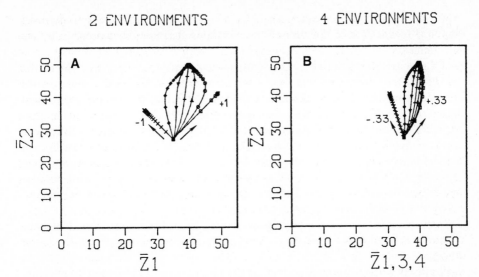

Fig. 7. Same as Fig. 6, but one environment (E2) is novel, and $q_1 = q_2 = 0.5$. *Symbols* are as in Fig. 6, and are marked every 50 generations. Initially, the mean phenotype is at (37, 27), with the optimal mean phenotype at (40, 50). Evolution occurs in the direction of the *arrows*. **A** Two environments. **B** Four environments. The optima in E1, E3, E4 are all 50 units, the optimum in E2 is 40 units

environment to overshoot the optimum, while a strongly negative genetic correlation can cause the phenotype in the old environment to drop below the optimum value.

Note, however, that unless the genetic correlation between the character states is ± 1, a population subject to a constant selective force will eventually attain the joint phenotypic optimum, and thus, the optimal level of phenotypic plasticity. When there is but a single phenotypic optimum, the constraints on the rate and direction of evolution that are imposed by a genetic correlation between character states are therefore only absolute if the correlation is ± 1. However, even a moderately high genetic correlation between character states can retard the process of evolution so much that a population can be held away from the optimum phenotype for a very long time.

Example 3: Four Environments, One is Rare

Now suppose that a population experiences four environments, each with a potentially different optimum phenotype. By extending the approach used in Example 1 in a highly symmetrical way, we can investigate whether selection in a larger number of environments can lead to greater limitations on the possibility of evolving an appropriate norm of reaction. For the multiple environments case, Eq. (2) can be modified to include the additional correlated responses. The dynamics of the character state expressed in the ith environment are then:

$$\Delta \bar{z}_i = q_i(\bar{W}_i/\bar{W})G_{ii}\nabla_i\ln\bar{W}_i + \sum_j q_j(\bar{W}_j/\bar{W})G_{ij}\nabla_j\ln\bar{W}_j. \qquad (4)$$

When there are multiple environments, Eq. (4) shows that the total correlated response to selection is the sum of the correlated responses that occur in all the environments.

To simplify this example of the multiple-environments case so that the results can be graphed in a two-dimensional plot (Fig. 6B), I assume that three of the environments have the same optimal phenotype (40 units), and are each represented as 28% of the total environment. The "rare" environment has a different phenotypic optimum (50 units) and composes only 16% of the total environment. Ecologically speaking, this is not a great difference in environmental frequency. Initially, the population is assumed to be roughly equally distant from the phenotypic optimum in all the environments. Finally, I assume that the total covariance across environments that we used in Example 1 is now divided among the three pairwise covariances between the "rare" and "common" environments.

Now, because 84% of the population is selected in the common environments, evolution is much more rapid there than in the rare environment. Evolution in the common environments has little effect on the response to selection in the rare environment when the genetic correlation produces a correlated response to selection in the direction of the optimum. However, for unfavorable correlations between the character states expressed in the rare and the common environments, the population can become so maladapted in the rare environment that it cannot maintain itself (i.e., the mean fitness of the portion of the population selected in the rare environment goes to zero). Extreme levels of maladaptation can occur even for relatively low values of the pairwise genetic correlations between the character states (Fig. 6B). Because this maladaptive effect can be so strong for even a slightly rare environment, one might expect it to be a potent force favoring the avoidance of rare environments in species that utilize a number of different environmental types. This could have important implications for evolution at species margins.

Example 4: Four Environments, One is Novel

Again, consider a very symmetrical situation. We postulate four environments at equal frequency, one of which is novel. The total covariance between the expression of a trait in the new environment and the other environments is the same as in Example 2, but now it is divided among the three "old" environments so that the pairwise correlations are much lower. Because 75% of the population is now selected in the "old" environment, the rate of adaptation to the new environment is considerably retarded (Fig. 7B). In this situation, the low individual correlations plus the slow rate of evolution causes the evolutionary trajectories to deviate less from a direct course to the joint phenotypic optimum than seen in Fig. 7A, although the rate of evolution is much slower than in the two-environment case. An interesting new property of the multiple-environments case arises here, however. If the pairwise correlations sum to produce a total correlation of -1, then the population will never attain the optimal phenotype even though no individual correlation is larger than -0.3 (Fig. 7B). This is an extension of an observation made by Dickerson (1955) that negative genetic correlations among n characters of $-1/n$ can halt evolution in a suite of traits expressed in a single environment.

Discussion and Conclusions

In order to study the evolution of phenotypic plasticity in a quantitative way, it is crucial to decide on a precise method for measuring plasticity that is amenable to evolutionary interpretation. A basic premise of this chapter is that the evolution of plasticity (the difference in the mean phenotype expressed in different environments) depends on two things: (1) The force of selection on plasticity is determined by the current mean phenotypic values exhibited by a population in each of several environments relative to some optimal phenotypic values for those environments (i.e., the current reaction norm relative to some optimal norm). (2) The response to this selection on the norm of reaction depends on the magnitude of genetic variation in the levels of phenotypic plasticity. Thus, we need a way to measure the mean and the genetic variance of phenotypic plasticity in a way that allows the dynamics of the response to selection of the phenotypic means expressed in different environments to be examined.

To measure these parameters, a character expressed in two environments is viewed as two character states, each of which is expressed in only one of the environments. Each character state has a mean phenotypic value in the environment in which it is expressed ($\bar{z}_i$, for the ith environment), and some genetic variance about that mean (G_{ii}). Moreover, the character states expressed in any two environments may genetically covary (G_{ij}, for $i \neq j$), such that the genetic correlation between the character states can be written as $G_{ij}\sqrt{G_{ii}G_{jj}}$.

In this framework, the difference between the phenotypic means in two environments is the mean level of phenotypic plasticity; the genetic correlation between the character states measures the extent of genetic variability in plasticity. Genetic correlations between states that are close to ± 1 correspond to the lowest genetic variation for a change in plasticity, while a genetic correlation between states of zero provides the highest variation for a change in plasticity.

The genetic correlation between character states expressed in different environments is statistically equivalent to the genotype-environment interaction (Robertson 1959; Fernando et al. 1984), a common measure of genetic variance in response to the environment. However, the genetic correlation between states is a more useful metric than is the g-e interaction for two reasons: (1) The magnitude of the g-e interaction increases monotonically as the genetic correlation between character states drops from $+1$ to -1 (Eq. 1). However, the available genetic variation for a change in the relative phenotype expressed in the two environments (a change in the current level of phenotypic plasticity) is at a maximum when the genetic correlation between states is zero because that is when the greatest potential exists for independent evolution of the states. (2) In contrast to the variance component due to g-e interaction, the genetic correlation between character states can readily be used in mathematical models that describe the dynamics of evolution of plasticity (e.g., Via and Lande 1985).

The genetic model described here is meant to assist in the formulation of reasonable hypotheses as to how various forces can influence the evolution of quantitative traits when populations experience stabilizing selection in several different environments. When the optimum phenotypes differ in the various environments populations undergo selection for phenotypic plasticity. Under most conditions, the models of evolution in a coarse-grained environment described here and in Via and Lande (1985) suggest that populations will eventually evolve the ability to adjust the phenotype appropriately under different environmental conditions. However, if the genetic correlation structure is unfavorable, this evolutionary process may take a very long time.

Several important points about the evolution of plasticity are illustrated by the numerical examples (Figs. 6, 7).

1. Populations are eventually expected to attain the optimal level of phenotypic plasticity, whatever it may be, *unless* there is
   a) No genetic variability for one or both of the character states.
   b) A pairwise genetic correlation between character states of ±1 (no genetic variation for the norm of reaction).
   c) A total genetic correlation between character states in a multiple environment situation of -1.
   d) No migration between the environments (no selection for the character state expressed in the alternate environment).
   e) A "cost" of plasticity, either metabolic or otherwise, that outweighs the selective advantage of phenotypic plasticity.
2. Any time that the mean character states expressed in the two environments are different, phenotypic plasticity will be observed (e.g., Fig. 6A). However, when the population is away from the joint phenotypic optimum, this plasticity will not correspond to a fully adapted state.
3. When the genetic correlation between the character states is unfavorable, the rate of evolution can be greatly retarded, keeping the population away from equilibrium for a long time. Thus, the plasticity that is observed in a population may not be adaptive. This possibility should be taken into account in field studies of phenotypic plasticity, and it underscores the need for direct estimates of the forces of selection in naturally occurring populations (e.g. Lande and Arnold 1983).
4. Due to the cumulative effects of genetic correlations with multiple environments (Eq. 4), populations may be limited in the total number of environments to which they can adapt simultaneously.

Even the simple models presented here may be useful to empiricists because they show that certain genetic and environmental parameters can have a great influence on the evolution of an appropriate phenotypic response to the environment. Estimation of phenotypic optima in different environments through explicit studies of natural selection (Lande and Arnold 1983) will permit the formulation of concrete hypotheses about what actually constitutes adaptive plasticity in particular cases and the extent to which populations are at optimum levels of plasticity.

Our models suggest that very different evolutionary trajectories can be generated by the same processes if parameter values are changed (e.g., cf. Figs. 6, 7). Thus, simply observing the dynamics of change in populations cannot provide enough information to determine the evolutionary mechanisms of phenotypic plasticity. However, by experimentally estimating the parameters used in the models (genetic variances within environments, genetic correlations across environments, and forces of selection), precise hypotheses about the future potential for evolution of plasticity in a given population can be formulated.

To date, few estimates of the genetic correlation between character states expressed in different environments are available for natural populations (Via 1984b: Rausher 1984; Weber 1985; Shaw 1986). As a larger data base of estimates of the critical parameters accumulates for natural populations, we may be able to test the hypotheses presented here and thus to further elucidate the evolutionary mechanisms of phenotypic plasticity.

Summary

Modification of the phenotype in response to different environmental conditions is commonly observed in both plants and animals. In many instances, such "phenotypic plasticity" appears to provide an advantage over the maintenance of a constant phenotype in different environments. Adaptive phenotypic plasticity is favored in populations in which there is natural selection toward different average phenotypic values in different environments and gene flow between them. The rate at which plasticity can evolve in such a situation is determined by the genetic variation in reaction norms that is available in the population. In this chapter, an experimental method for estimating the magnitude of genetic variation for phenotypic plasticity is described. This method involves considering a character expressed in several environments as a set of genetically correlated *character states,* each of which is expressed in only one of the environments; the appropriate character states are then measured on family members that have been allowed to develop in the various environments. A quantitative genetic model of evolution in a patchy environment is presented that employs the genetic correlation between character states as one of the parameters. Using this mathematical framework, the effects of different patterns of genetic variation in phenotypic plasticity on the evolution of plasticity can be determined. Such genetic models of evolution are useful to empiricists because they reveal genetic and ecological parameters that can influence the evolution of phenotypic plasticity and they suggest how experimental estimates of these parameters can be interpreted.

Acknowledgment. My thanks to Russell Lande for all his help during the formulation of these ideas. Cindy Norton, Dan Thompson, and Andy Clark provided useful criticisms of the manuscript. This work was supported by USPHS National Research Service Award GM09593, NIH grant GM34523, NIH grant GM27120 (to R. Lande), and a Searle Scholars Award from the Chicago Community Trust.

References

Atkinson BG, Walden DB (eds) (1985) *Changes in Eukaryotic gene expression in response to environmental stress.* Academic Press, London

Bernays E (1986) Diet-induced head allometry among foliage-chewing insects and its importance for graminivores. Science 231:495–497

Bradshaw AD (1965) Evolutionary significance of phenotypic plasticity in plants. Adv Genet 13: 115–155

Briggs D, Walters SM (1984) *Plant variation and evolution.* 2nd ed. Cambridge University Press, Cambridge

Clough JM, Teeri JA, Alberte RS (1979) Photosynthetic adaptation of *Solanum dulcamara* L. to sun and shade environments. I. A comparison of sun and shade populations. Oecologia (Berl) 38:13–22

Dickerson GE (1955) Genetic slippage in response to selection for multiple objectives. Cold Spring Harb Symp Quant Biol 20:25–32

Falconer DS (1952) The problem of environment and selection. Am Nat 86:293–298

Falconer DS (1981) *Introduction to quantitative genetics.* 2nd ed. Longman, New York

Fernando RL, Knights SA, Gianola D (1984) On a method of estimating the genetic correlation between characters measured in different experimental units. Theor Appl Genet 67:175–178

Freeman GH (1973) Statistical methods for the analysis of genotype-environment interactions. Heredity 31:339–354

Gause GF (1947) Problems of evolution. Trans Conn Acad Sci 37:17–68

Gilbert JJ (1966) Rotifer ecology and embryological induction. Science 151:1234

Harvell CD (1984) Predator-induced defense in a marine bryozoan. Science 224:1357–59

Hochachka PW, Somero GN (1984) *Biochemical adaptation.* Princeton Univ Press, Princeton, NJ

Jain SK (1978) Inheritance of phenotypic plasticity in soft chess, *Bromus mollis* L. (Gramineae). Experientia (Basel) 4:835–836

Khan MA, Antonovics J, Bradshaw AD (1976) Adaptation to heterogenous environments. III. The inheritance of response to spacing in flax and linseed (*Linum usitatissimum*). Aust J Agric Res 27:649–659

Kuhlmann HW, Heckmann K (1985) Interspecific morphogens regulating prey-predator relationships in protozoa. Science 227:1347–1349

Lande R (1979) Quantitative genetic analysis of multivariate evolution, applied to brain: body size allometry. Evolution 33:402–416

Lande R (1980) The genetic covariance between characters maintained by pleiotropic mutations. Genetics 94:203–215

Lande R, Arnold S (1983). The measurement of selection on correlated characters. Evolution 37:1210–1226

Levins R (1968) *Evolution in changing environments.* Princeton Univ Press, Princeton, NJ

Nijhout HF, Wheeler DE (1982) Juvenile hormone and the physiological basis of insect polymorphisms. Q Rev Biol 57:109–133

Orzack SH (1985) Population dynamics in variable environments V. The genetics of homeostasis revisited. Am Nat 125:550–572

Prosser CL (1973) *Comparative animal physiology.* Saunders, Philadelphia

Rausher MD (1984) Tradeoffs in performance on different hosts: Evidence from within and between site variation in the beetle, *Deloyala guttata.* Evolution 38:582–595

Robertson A (1959) The sampling variance of the genetic correlation coefficient. Biometrics 15:469–485

Scheiner SM, Goodnight (1984) The comparison of phenotypic plasticity and genetic variation in populations of the grass *Danthonia spicata.* Evolution 38:845–855

Schmalhausen II (1949). *Factors of evolution: the theory of stabilizing selection.* Blakiston, Philadelphia

Schlichting CA (1986). The evolution of phenotypic plasticity in plants. Annu Rev Ecol Syst 17:667–693

Schlichting C, Levin LA (1984) Phenotypic plasticity of annual phlox: tests of some hypotheses. Am J Bot 71:252–260

Shaw RG (1986) Response to density in a wild population of the perennial herb *Salvia lyrata:* variation among families. Evolution 40:492–505

Teeri JA (1978) Environmental and genetic control of phenotypic adaptation to drought in *Potentilla glandulosa* Lindl. Oecologia (Berl) 37:29–39

Turelli M (1984) Heritable genetic variation via mutation-selection balance: Lerch's zeta meets the abdominal bristle. Theor Popul Biol 25:138–193

Via S (1984a) The quantitative genetics of polyphagy in an insect herbivore. I. Genotype-environment interaction in larval performance on different host plant species. Evolution 38:881–895

Via S (1984b) The quantitative genetics of polyphagy in an insect herbivore. II. Genetic correlations in larval performance within and across host plants. Evolution 38:896–905

Via S, Lande R (1985) Genotype-environment interaction and the evolution of phenotypic plasticity. Evolution 39:505–523

Via S, Lande R (1987) Evolution of genetic variability in a spatially heterogeneous environment: effects of genotype-environment interaction. Genet Res (in press)

Weber G (1985) Genetic variability in host plant adaptation of the green peach aphid, *Myzus persicae.* Entomol Exp Appl 38:49–56

Wheeler DE, Nijhout HF (1981). Soldier determination in ants: New role for juvenile hormone. Science 213:361–363

Yamada Y (1962) Genotype x environment interaction and genetic correlation of the same trait under different environments. Jpn J Genet 37:498–509

Chapter 4 Reflections on the Genetics of Quantitative Traits with Continuous Environmental Variation

A. J. van Noordwijk and M. Gebhardt[1]

Introduction

The process of evolution is complex. The number of variables that can be studied simultaneously is therefore too limited to capture the whole process. The usual simplifications that are made by geneticists, ecologists, developmental biologists and paleontologists exclude the variables studied in the other disciplines. However necessary these simplifications are, they lead to a neglect of the important inter-actions between processes. In this chapter we explore some aspects of the quantitative genetic methods that are ecologically unrealistic. In particular, we consider one or more continuously varying environmental parameters, such as temperature and food, that are either controlled or uncontrolled, and ask how the effects of these environmental parameters on single and on correlated traits can be studied.

Although we mainly discuss general problems, it is good to keep particular biological systems in mind as a reference. We will switch our thinking between *Drosophila, Daphnia* and *Parus*. *Drosophila* is a general genetic workhorse, for which it is possible to create ecologically more interesting conditions in the lab than is often realized. A cyclical parthenogenetic animal such as *Daphnia* is an excellent reference system for problems where one needs many copies of the same genotype to obtain an efficient experimental design. Our third reference system is a passerine bird such as the Great Tit (*Parus major*), where ringing allows the compilation of family trees and therefore quantitative genetic analysis in a natural environment for a species with a well-known ecology.

It is one of the axioms of quantitative genetics that the environment affects the traits studied and that therefore all estimates are local to the particular environment and the particular population from which the data were obtained. In practice, this very restricted nature of quantitative genetic parameters is somewhat relaxed through the assumption that the population and the environment studied are representative for a much larger group, and that therefore the estimates are also representative for this wider group.

The same axiom has, however, led to another restriction in virtually all studies, namely that the environmental variation represented in the study has been minimized. This not only has the practical advantage of greater resolution of the genetic effects and greater reproducibility, but it also eliminates several potential complications

[1]   Zoologisches Institut, Rheinsprung 9, CH 4051 Basel, Switzerland

Genetic Constraints on Adaptive Evolution
Ed. by V. Loeschcke
© Springer-Verlag Berlin Heidelberg 1987

in the analysis. Furthermore, in the agricultural setting of quantitative genetics it does not make sense to create a less than (profit) optimal environment. In its turn, this practice has led to the implicit assumption that the environmental variation left in the experimental situation is uncontrollable noise, i.e. error variance. If environmental variance is dealt with explicitly, it is nearly always in the form of discrete environments. This allows one to define the same trait in different environments as two different traits, and to calculate genetic correlations between these traits and similar trait-environment traits (see Via, this Vol. for an example). There are cases where this representation provides a good description of a natural situation, such as an insect with two species of host plants where the larvae either grow up on species A or on host B.

From an ecological point of view, however, almost all environmental variation is continuous rather than discrete. Moreover, if one takes the phenotypic variation in a natural population as the starting point, a substantial portion can usually be explained from the variation in environmental parameters, as ecophysiologists continuously demonstrate. Egg size in the Great Tit was shown to have a heritability of about 0.7 (van Noordwijk et al. 1981). In the same data a strong correlation between egg size and the temperature in the 3 days preceding the laying of the egg is present as long as the temperatures remain below $12^{\circ}C$ (we will return to this example below). This leads to correlations between temperature in the laying season and repeatability for egg size. It is clear that heritability estimates also depend on the temperatures in the seasons when the data were collected. This raises several questions, such as: Were the temperature conditions representative for this population?

It also raises the possibility that an overall estimate of heritability gives a false prediction of genetic change due to selection, for example when selection for body size is especially strong under poor food conditions; heritability is absent under those conditions. Using an average heritability and an average selection pressure would then give a false impression. Thus, from an ecological point of view, the scale – both spatial and temporal – is critically important. In a variable environment the time when an environmental factor affects the phenotype also becomes an important parameter.

In this chapter we explore the possibilities of studying quantitative genetics in environments with considerable variation. We start with a description of reaction norms and how the variation along the reaction norm is represented in the data if the environmental factor is not controlled. We will then expand into two environmental axes and two traits. Finally, we will present data from an experiment where reaction norms for development time and size at eclosion were measured in *Drosophila*.

Reaction Norms

Woltereck (1909) defined the reaction norm as the set of all phenotypes that are
displayed by a single genotype under different environmental conditions. He worked
with *Daphnia* and was interested in body size and the development of helmet-shaped
heads that are observed during part of the year. The helmet-shape and other spines
are effective in reducing invertebrate predation and their formation is induced by a
chemical produced by these predators (Hebert and Grewe 1985). There are differences between clones in the inducibility of these spines. These differences are
correlated with the presence of predators in the natural environment (see Hebert
1984 for a recent brief review and references). These aspects of cyclomorphosis
make it a paradigm for reaction norms.

For a single environmental factor, the reaction norm can be seen as a function
of the environment that gives the phenotype (see Schmalhausen 1949). These functions may have any form, ranging from a steplike response above a certain threshold
to a gradual change in phenotype (see e.g. Rendel 1959; Scharloo 1962; Scharloo,
this Vol.). Different genotypes may then have different functions. This representation is frequently used to illustrate the fact that the presence of genetic variation
depends on the environment in which the observations are made. Although reaction
norms are thought of as functions of a continuously varing environmental parameter,
they are, of course, estimated from measurements at discrete values over a range
of the environmental variable. In Fig. 1, three hypothetical reaction norms are given
(for other examples see Clark, this Vol. and Scharloo, this Vol.). Together with

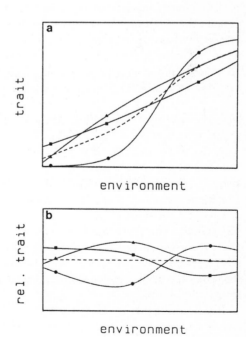

Fig. 1. The reaction norms of three genotypes are plotted together with the mean
phenotype for every value of the environmental parameter in **a**. In **b** the reaction
norms of the same three genotypes are now
plotted relative to the population mean

the three reaction norms, the mean phenotype of these three genotypes has been drawn. In quantitative genetics it is customary to express all observations relative to the population mean. One can then redraw the reaction norms as deviations from the population mean (Fig. 1b).

In a classical analysis of the performance of several genotypes over a range of environmental conditions, it is possible to combine different environmental factors on a single axis (see Falconer 1981). One characterizes every environment by the mean value of all the tested genotypes. One can then characterize every genotype by the regression of phenotypic values over the range of environments. This procedure carries the implicit assumption that the mean phenotypic value in the same environment is the relevant mean. In ecological terms this would imply that organisms only encounter individuals that grew under the same environmental conditions. This might be appropriate for a culture of bacteria, but it seems unrealistic for most animals and plants. In most natural populations the individuals that compete with each other will also differ in phenotype due to variation in environmental conditions, especially when it is the environmental conditions at some moment in the past, e.g. during growth, larval stage, seedling stage, that are important in shaping the phenotype. We discuss this topic further below. The relevant point here is that the amount of environmental variation that should be taken into consideration depends on the interplay with the population structure.

Known and Unknown Environmental Conditions

In laboratory experiments one normally tries to minimize the uncontrolled environmental variation. How can one exptrapolate from laboratory measurements of reaction norms for temperature, for example, to a natural population living in an environment with substantial daily and yearly temperature fluctuations? The environment can, at least in theory, be characterized by a frequency distribution of temperature conditions. When the reaction norms of particular genotypes are known, one can construct a frequency distribution of phenotypic values of each genotype from the frequency distributions of the environmental conditions. This is illustrated in Fig. 2.

A figure like this can most easily be obtained for a species that can be cloned. In Fig. 2 we have the reaction norms for three genotypes, which lead to the distributions of phenotypic values given in Fig. 2c through 2e under the environmental conditions given in Fig. 2b. The variance of these distributions of phenotypic values depends on the slope of the reaction norm and the variance of the environmental distribution. These are theoretical phenotypic distributions that do not contain any environmental variance due to other factors. Apart from a construction using reaction norms and environmental frequency distributions, it could be measured as the difference in the environmental variance in a constant and in a variable environment within a clone.

It seems attractive to divide the total environmental variance in a component (f) due to a known factor, that could be controlled, and a component (e) due to other

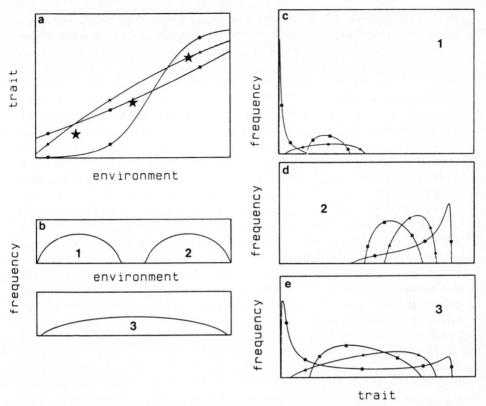

Fig. 2. In **a** the reaction norms for three genotypes are given. The *stars* indicate the population means at environmental values corresponding to the means of the three distributions of environmental values given in **b**. These distributions will lead to the distributions of phenotypic values for the three genotypes that are given in **c**, **d** and **e** respectively. Note the bimodality of the distribution of the genotype marked by *solid circles* in **e**, which is due to the stretching of the phenotype axis in the range of the steep incline of the reaction norm

factors, together with the genetic variation that is independent of environment (g). We can write $p = g + e + f$, where the phenotypic value (p) is the sum of these three components. At the population level we obtain:

$$V_p = V_g + V_f + V_e + 2\,COV_{gf} + 2\,COV_{fe} + 2\,COV_{ge} \tag{1}$$

When the component f is zero, this is the usual expression. In this model, g, e, and f are assumed to be independent. In the context of reaction norms, however, the interactions between the components are of special interest. Ignoring the covariances, we would obtain:

$$V_p = V_g + V_f + V_e + V_{Igf} + V_{Ige} + V_{Ife} + V_{Igef} \tag{2}$$

A non zero V_{Igf} implies genetic variation in the reaction norm studied, while V_{Ige} stands for variation in developmental stability between the genotypes. When the developmental stability, that is the amount of environmental variance, depends on the systematically varied environmental factor this is found in the V_{Ife}. These interaction variances can be obtained from a factorial ANOVA in which several genotypes are treated with a range of one environmental factor and different amounts of uncontrolled environmental variance. Alternatively, one could equate the uncontrolled environmental variance with the error variance, but this would not allow one to detect interactions with e.

This model is more general than the one used by Scheiner and Goodnight (1984, and refs therein). Their aim is to separate the genotype-environment interaction component. Their interaction component is equivalent to our component f plus the interaction variances. We have chosen this formulation because it emphasizes that whether or not one treats part of the environmental variance more explicitly is entirely a matter of analysis.

The advantage of this approach is that it allows any distribution of the systematically varied environmental factor, and also that it allows investigation of situations where the environmental factor is not always precisely known, but nevertheless varies. For example, the effects of temperature on egg size in the Great Tit were determined in a special study with intense monitoring (see van Noordwijk 1984). It was found (see Fig. 3) that at temperatures below 12°C, a positive correlation existed between temperature in the preceding 3 days and the size of the egg. This affects the repeatability and heritability estimates. This is most easily seen in repeatability estimates for egg volume that include the size of eggs produced in first and in second broods. Data for 3 years are given in Table 1, together with the mean temperatures during the laying periods (van Noordwijk et al. 1981; Keizer et al., in prep.). In principle, it would be possible to correct for the effect of temperature, i.e. remove the component f from our equation, to see if the repeatability estimates are otherwise similar. This is necessary, because it is often practically impossible to obtain sufficiently large sample sizes from the data from a single season, or from a single experiment.

One interesting problem is that it is difficult to find a suitable reference against which to express the temperature effects on egg size. A first thought is to use the

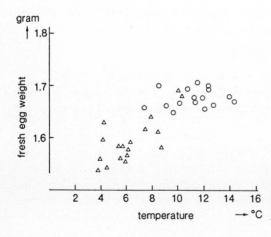

Fig. 3. The mean size of all eggs laid on a particular day, as a function of the average ambient temperature in the preceding 3 days. ○ Data for 1978; △ Data for 1979; r = 0.74 (N = 18, P < 0.01) from van Noordwijk (1984)

Table 1. Repeatability estimates for egg volume, across first and second broods in relation to the mean temperatures during egg laying (Data from the Great Tit population on Vlieland, The Netherlands; van Noordwijk et al. 1981 and Keizer unpubl.)

Year	n (indiv.)	Repeatability	T in first	T in second
1977	18	0.72	9.3	13.7
1978	20	0.39	8.2	13.2
1979	31	0.09	7.1	13.0

mean egg size in a clutch as a reference point. However, this mean egg size depends on the temperatures under which the other eggs in the clutch were laid. An egg laid at a temperature of 9°C can be relatively small if 9°C was the coldest day during the laying of the clutch, or it can be relatively large if 9°C was relatively warm. The variation that might exist between females in their reactions to temperature could be investigated by comparing the change in temperature from one day to the next with the change in egg size (van Noordwijk 1984). It was found that at shifts in temperatures of about 2°C, nearly all individuals reacted in the same way.

This example shows how difficult it can be to find a satisfactory reference point. It strongly suggests that even when reaction norms cannot be measured as such, they can be very helpful as a guide to design the best parametrization of the variables.

Selection for a Single Trait

The representation with the phenotypic frequency distribution of a given genotype may also be used to look at selection from a different angle. Let us assume, as is normally done, that selection acts on the phenotypic value for a particular trait. In a population consisting of many individuals belonging to a few genotypes (clones), the operation of selection is illustrated in Fig. 4. The width of the distributions relative to the distances between the means of the distributions is a measure of the heritability of the trait, a high heritability being obvious as a relatively large difference between clones.

Together with a fitness function (Fig. 4b), which expresses relative fitness depending on phenotype, frequency distributions after selection may be constructed (Fig. 4c). These can be used to calculate the relative frequencies of the clones in the next generation. The shapes of the distributions of phenotypes of each clone remains the same as it was in the previous generation if the environment is the same, because all the phenotypic variation within a clone is due to environmental variance (Fig. 4d). With clones, the operation of selection can be summarized in the changes in frequency of the different clones.

In a crossbreeding species, the operation of selection is similar. It is, of course, impractical to draw phenotypic frequency distributions for each genotype. In theory, however, one can regard every individual as a sample (size 1) of a distribution similar

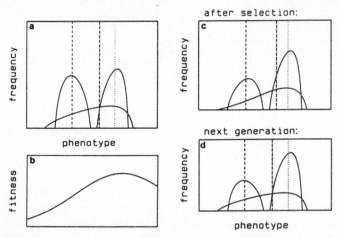

Fig. 4. In **a** the phenotypic distributions of three genotypes are depicted. Assuming the normal fitness function shown in **b**, the phenotypic distributions shown in **c** will result by selection. Note the shift of means (*vertical lines*) and the changes in shape. In **d** the phenotypic distributions in the next generations are shown. The shapes and locations are the same as in **a** (one generation earlier, before selection), but the areas (relative occurrences of genotypes) have changed

to that for clones. One can then look at selection either as a deterministic process on the phenotype having a stochastic genetic component, or (as is done here) first combine the environmental deviation with the fitness function and then treat the selection on the genotype as a deterministic process. This latter approach is similar to the usual treatment of major genes, where deterministically operating selection coefficients are used. Unfortunately, it is not customary to regard the selection coefficients as explicit functions of environmental variables.

One Trait with Two Environmental Factors

There is little difference between the phenotype distribution of a single genotype due to a known and due to an unknown environmental factor. Let us assume that we have two environmental factors that are uncorrelated and for which the phenotypic responses are also independent. This latter assumption is unrealistic, but the former can be created in an experiment, using e.g. temperature and food quality. In measuring the reaction norm for one variable, the other variable also has a certain value. If this other variable is not fully controlled, it leads to a distribution of phenotypes like that depicted in Fig. 2. This is shown in Fig. 5, where the reaction norm is given in the form of successive phenotypic distributions. The shapes of the distributions depend on the value of the other environmental factor. In Fig. 5a, we have used the distributions from Fig. 2d, and those from Fig. 2e were used in Fig. 5b. The mean values for each of the distributions would give us the reaction norms as they are

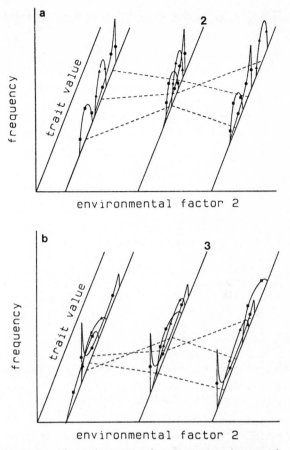

Fig. 5. a A three-dimensional representation of reaction norms for a second environmental factor, if measured under distribution 2 of the first environmental factor in Fig. 2. **b** The same if measured under distribution 3 in Fig. 2. In both cases it is assumed that the reaction norms in Fig. 2 were measured under fixed conditions corresponding to the middle value of this second environmental parameter. The means of each of the distributions would give the reaction norm as it is usually measured. The *broken lines* therefore give the reaction norms for factor 2. The figure shows how the precision of measuring reaction norms can be dependent on other environmental variables

usually measured. Interactions between the environmental variables would become visible as changes in shape of the phenotypic distributions of the single clones.

It is possible to use this representation of the effects of two environmental factors as a basis for an analysis of selection. This would require a bivariate fitness function and bivariate frequency distributions of environmental conditions.

Two Traits, Complex Traits and a Single Environmental Factor

So far we have regarded single traits in complex environments. In many cases, it is more interesting to study complexes of traits. It is, of course, possible to determine the reaction norms for two or more traits simultaneously. An example is given in Fig. 6. In the first two frames, the phenotype is plotted against the value for the environmental factor. This is the perspective that an ecophysiologist would have. It is not clear from these representations whether there is any relation between the two traits.

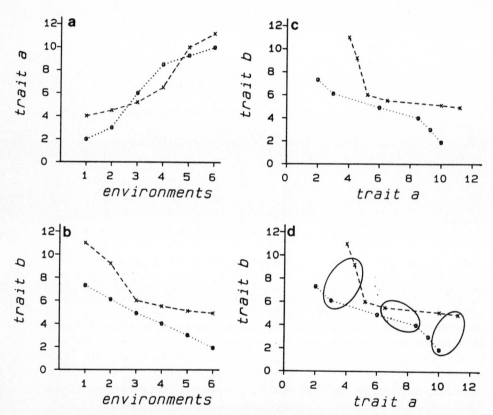

Fig. 6 a–d. The phenotypic values for two genotypes are given along the same environmental gradient for two different traits in **a** and **b**. For the purpose of reference six equidistant environmental values are indicated. **c** A scatter plot of these two traits where the points represent one of the two genotypes under one of the six environmental conditions. The bivariate reaction norms, connecting the phenotypes of a single genotype have also been drawn. If these two genotypes are extremes and other genotypes in a population are simultaneously intermediate for both traits, then envelopes around these points give an indication of the correlations between the traits. In this example, negative correlations are observed in environments 3 and 4 and in estimates across all environments, whereas positive correlations are found in the other four environments

It may be especially insightful to draw the reaction norms in a scatter plot of the two traits as in Fig. 6c (see Stearns 1983; Stearns and Koella 1986). If this is done for several genotypes, one may get an impression of the correlation structure by drawing an envelope around all the points obtained in a single environment. This is shown in Fig. 6d. These correlations are the correlations between the genotypic values and give an impression of the genetic correlation; they lack the scaling by the heritabilities.

In this hypothetical example (as in our real example, see Fig. 7 below) there are several positive and several negative correlations, depending on which environment is regarded. This is due to the crossing of reaction norms and/or differences in the location of the changes in the slope of the reaction norms. It is impossible to make any quantitative statements, but we were surprised by the ease with which physiologically plausible reaction norms led to sign changes in the correlations between traits over a range of environments.

A Case Study

Reaction norms of developmental time and of weight at eclosion in *Drosophila mercatorum* were measured over a range of yeast concentrations. The data were collected during pilot experiments to get a first impression of the extent of genetic variation in a program aimed at selection for reaction norms of compound life-history traits.

Methods

A more detailed description of the experimental procedure is given elsewhere (Gebhardt et al., in prep.). The experiments were carried out with the F1 from crosses between males of six different isofemale strains and females from two isozygous strains of *D. mercatorum*. The isofemale strains were derived from single fertilized females which were collected by A. Templeton at three sites near Kamuela (Hawaii). The isozygous strains also stemmed from Kamuela populations, but were much older. They were derived by selection for parthenogenetic reproduction, were highly efficient for this mode, and corresponded to the strains maintained by A. Templeton labelled K280Im and K23aFImSF2. Because all the females within an isozygous strain are genetically identical, the experimental design thus comprises essentially two families of half-sibs, corresponding to the two isozygous strains, barring maternal effects due to different environments experienced by the individual females.

Different environments were created by adding different amounts of dried yeast to a standard cornmeal/sucrose/agar medium yielding yeast concentrations of 0.1, 0.25, 0.5 and 1.5%. Freshly emerged females and males were put together for 4 days. Thereafter, females were allowed to oviposit during 12 h on a substrate having the same yeast concentration as experienced later by the larvae. Larvae were collected 24 h after the end of egg laying, by adding a concentrated sucrose solution which

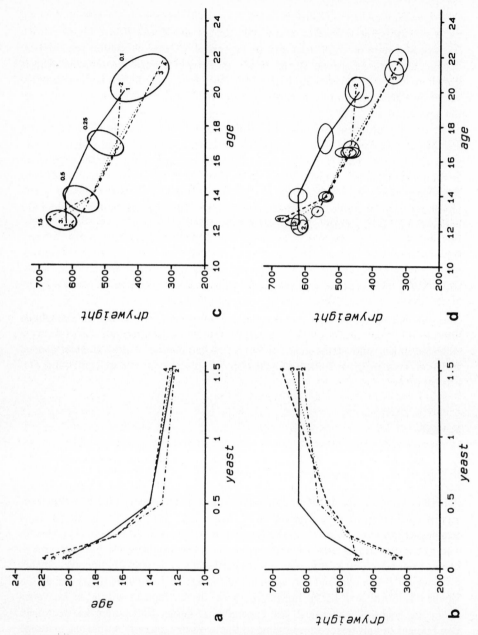

Fig. 7 a–d. Reaction norms of four genotypes along a gradient of yeast concentrations (% dried yeast in the culture medium). Genotypes *1* and *2* versus *3* and *4* correspond to the two iso-zygous maternal lines and constitute half-sib families. **a** Age at eclosion in days. **b** Dry weight at eclosion in micrograms. Note the reversal of genotypic ranks between the extreme yeast concentrations. **c** Dry weight plotted against age. Corresponding yeast concentrations are enclosed by *ellipses*, the concentration is indicated by the *number*. Note the changing sign of the genetic correlation between the two traits. **d** Same as c, with the difference that *ellipses* represent 95% confidence regions for each genotype x environment combination

caused them to rise to the surface. They were transferred to tubes containing 2 ml of the experimental medium and kept under constant conditions of $25 \pm 0.1^\circ C/70\%$ rel. humidity until the last adults had emerged. These were collected at 12 h intervals, sexed and dried for 4 h at $70^\circ C$ for determining their dry weight. For each cross and treatment combination there were between five and ten replicate tubes containing ten larvae each.

Results and Discussion

A lower nutritional value of the culture medium affected all the flies in a qualitatively similar manner as illustrated by Fig. 7 a–d. We only show the results for four of the twelve crosses. The other crosses were similar and usually intermediary between the ones shown. Development was substantially delayed (Fig. 7a) and the emerging flies had a lower dry weight (Fig. 7b) irrespective of genotype. Moreover, phenotypic variability was substantially increased at lower yeast concentrations (Table 2). An ANOVA performed on the mean values from replicate tubes revealed highly significant effects of yeast and genotype for dry weight and log-transformed age and of the interaction yeast*genotype for dry weight (Table 3).

Closer inspection of Fig. 7 shows a genotype-environment interaction for both traits in that differences between genotypes are more pronounced at lower yeast concentrations. Moreover, the reaction norms for the dry weight of genotypes 1 and 2 intersect with those of genotypes 3 and 4 (Fig. 7b), producing a reversal of the correlation between the two traits along the yeast gradient (Fig. 7c and d, Table 4). It should be mentioned that genotypes 1 and 2 versus 3 and 4 represent natural groups in that they correspond to the two maternal lines used for producing the crosses. Flies within either of these groups are thus genetically half-sibs and should be more similar than between these groups. This is evident at the extreme values of the yeast gradient. At intermediate values, all differences disappear due to the crossing of the reaction norms for dry weight.

Table 2. Means and standard deviations for age and dry weight at eclosion for female flies, genotypes pooled [a]

Yeast		0.1		0.25		0.5		1.5
Age:	Mean	21.0	>>>	17.0	>>>	13.7	>>>	12.5
	SD	2.8	>>>	1.7	>>>	1.1	>>	0.7
Dry weight	Mean	379	<<<	483	<<<	562	<<<	652
	SD	108	>>>	82	>>>	64	>	56

[a] Age in days, dry weight in micrograms. Levels of significance were determined using non-metric tests ($>>> : P < 0.001, >> : P < 0.01, > : P < 0.05$)

Table 3. ANOVA for the effects of yeast and genotype[a]

Source	df	MS: Log(age)	MS: dry weight
Yeast	3	4.551 ***	1.075 ***
Genotype	9	0.017 ***	0.031 ***
Yeast*genotype	27	0.006 n.s.	0.010 ***
Error	237	0.006	0.003

[a] The means from replicate tubes were used as data. All significances (***) indicate $P < 0.001$

Table 4. Estimates for genetic variances, covariance and correlation of age and dry weight, based on the means over crosses

Yeast	Var(age)	Var(dry weight)	Cov	r	P
0.1	1.14	3882	−57.4	−0.86	0.0007
0.25	0.96	1509	12.1	0.32	n.s.
0.5	0.11	916	5.4	0.53	n.s.
1.5	0.08	680	5.6	0.76	0.007

Figures 7 a–d were constructed using the means of the crosses. Barring an interaction between the genotypes and uncontrolled environmental influences, the relationships visible in this figure should thus approximate genetic relationships, because the phenotypic variations induced by the uncontrolled environmental factors are diminished by a factor corresponding to the sample size (between five and ten replicate vials or between 40 and 50 individuals in most cases). With the necessary qualifications in mind, we may thus take this case as an example for changes in the heritabilities and the genetic correlation structure due to genetic differences in the response to an environmental factor.

The Use of Optimality Models

We have outlined a conceptual framework for analyzing the genetic aspects of adaptive phenotypic plasticity. By measuring reaction norms for several traits along different controlled environmental axes and for a number of genotypes, it should be possible to get an impression of the genetic constraints that have impact on the dynamics of the evolution of phenotypic plasticity. Such constraints can be thought of as heritabilities differing for different traits and for different environments, pleiotropic effects of genes on different traits, pleiotropy being particular for different environments and thereby giving rise to a non-constant genetic correlation matrix of traits among environments. These and other complications can be recognized as consequences of the genotype-specific reaction norms.

Optimality models can be used to predict the shape of reaction norms within a specified framework of physiological constraints and incorporating age and environment specific mortalities. These models usually do not consider the constraints imposed by the developmental system and the genetics governing the development of the traits of interest (Stearns and Koella 1986). In investigating the role of genetic constraints, it can be very helpful to know what would be expected without them. Especially if several interacting traits are considered simultaneously in interaction with the environment, it becomes impossible to predict the optimal phenotype without explicit modelling.

The optimal phenotype is here used as a summary of a fitness curve, by giving its mode. We therefore disagree with Rose (this Vol.) that optimality models tend to be vacuous when explicit genetics are absent from them. We agree fully with Rose (ibid.) that a selection experiment designed to test a theory based on an optimality model could fail, although the underlying assumptions about physiology and mortality schemes had been checked and shown to be correct. The reason would be that the predicted solution was not within the set of possible phenotypes defined by the genetics of the developmental system.

The analysis of reaction norms yields insight into the genetic constraints at work. Changing heritabilities and correlations of traits along an environmental axis can immediately be understood if the reaction norms of different genotypes are not parallel, especially if they cross. The selection on reaction norms depends on the frequency with which the different environments are encountered. In principle, it should be possible to complement a theory of optimal reaction norms like that proposed by Stearns and Koella (1986) with the above theory of genotype x environment interaction to derive predictions about the evolutionary dynamics of phenotypic plasticity.

The Physiological Basis of Pleiotropy

Antagonistic pleiotropy is a potentially important genetic constraint to adaptive evolution (see Clark, Rose et al., Via, all in this Vol.). We will not repeat what they have said, but provide a somewhat different perspective to the same problem.

The physiological processes that lead to a given trait are relevant for understanding the interaction between genes and environmental influences in at least two ways. The first important aspect is timing. This information is important in judging at which time the environmental conditions are important. This may be different for the same trait in different species.

If we look at egg size in birds, for example, physiologists make a distinction between income breeders and capital breeders, depending on whether the resources put into the egg are collected in a short period before laying that egg (income breeders) or whether the bird (e.g. a goose) collects resources during spring migration and produces the entire clutch out of stored energy reserves (Thomas 1987). In the analysis of the effects of temperature on egg size in the Great Tit (see above) we used the fact that it is an income breeder and that the within-clutch variation in egg size reflects a large part of the environmental variance for egg size. In a typical capital breeder, one expects all eggs in a clutch to be similarly affected by the conditions during the buildup of the fat reserves. In thise case, trade-offs between egg size and number of eggs are of course more likely than in the case of income breeders.

This brings us to the second important aspect of the physiology. To what extent do the physiological/developmental mechanisms leading to two traits overlap? One can regard most physiological processes as an allocation of energy to a particular process. Especially in life-history theory the allocation of resources to different traits is a general metaphor. This makes it possible to view the relations between any two traits as Y-shaped. There is a shared part where energy flows from the environment into both traits, and a point where the physiological pathways diverge, where an allocation has to be made to either of the two traits. Different phenotypic values between individuals will arise somewhere along the whole physiological pathway, that is somewhere between the acquisition of the resources and the measurable phenotype.

Irrespective of the source of these differences in physiology, whether genetic or environmental, when they occur in the common part of the energy pipeline leading to the two traits, they will contribute to a positive correlation between the two traits. When they affect the junction between the energy flow leading to the two traits, they will contribute to a negative correlation between the two traits. The net result for a correlation between two traits depends on the relative amounts of variation in the common and the separate parts of the physiological pathways (see van Noordwijk and de Jong 1986).

A description of the whole developmental and physiological network as a pipe with one entrance and two exits is, of course, a gross oversimplification. Nevertheless, some aspects of the pipeline metaphor are also true for much more complicated structures. Depending on which link in the chain of reactions is altered, the effect on two traits will vary in the same or in opposite directions. Thus, a generally made assumption that genetic correlations between traits will be qualitatively different from the phenotypic (or the environmental component therein) correlation implies that environmentally caused variations occur at different places in the network compared to genetically caused variations. Whether this is true or not depends on the genetic variation present, but also on the amount and sort of environmental variation present. It is argued elsewhere (van Noordwijk, in prep.) that the conditions under which the genetic correlation will be different from the phenotypic one are much more likely in laboratory and in agricultural populations than in most natural populations.

Here, the point is that the importance of negative pleiotropy depends on the environmental variation present in the study. Pleiotropic effects of genes are important in our understanding of genetic constraints to adaptive evolution. If one accepts that the nature of the environmental variation present in a study can have a substantial effect on the differences between genetic and phenotypic correlations between traits, it follows automatically that it is dangerous to draw strong conclusions on the importance of negative pleiotropy as a genetic constraint on data gathered from populations in an ecologically unrealistically simple environment. This does not necessarily mean that data have to be collected in the field (cf. Clark, this Vol.); it is quite possible to create interesting environments in the laboratory even for *Drosophila*. A change in the relation between genetic and environmental correlations may either be brought about through an effect of the environment on the environmental correlation, or on the genetic correlation. In the *Drosophila* data above, the genetic component in the correlation changes its sign depending on the environment.

Conclusion

The analysis of quantitative traits in more natural, environmentally more structured, environments is just beginning. It is already apparent in a number of case studies that the relation(s) between the environmental conditions affecting the phenotype and the environment determining the forces of selection, are far from simple. In this contribution we have sketched a framework for dealing with these problems. We will need both many case studies of the genetics in structured environments and theories that integrate ecology, genetics and developmental biology to understand which genetic constraints are of quantitative importance in the process of evolution. We therefore believe that insight into the quantitative genetics in structured environments is a necessary precondition for the understanding of genetic constraints to adaptive evolution.

Acknowledgments. We thank G. de Jong, P.H. van Tienderen and S.C. Stearns for helpful discussions. We thank the Swiss Nationalfonds for financial support (grant 3.131-0.85 to A.J. v. N. and grant 3.642-0.84 to S.C. Stearns).

References

Falconer DS (1981) Introduction to quantitative genetics, 2nd ed. Longman, London
Hebert PDN (1984) Demographic implications of genetic variation in zooplankton populations. In: Wöhrmann K, Loeschcke V (eds) Population biology and evolution. Springer, Berlin, Heidelberg New York, pp 195–207
Hebert PDN, Grewe PM (1985) *Chaoborus*-induced shifts in the morphology of *Daphnia ambigua.* Limnol Oceanogr 30:1291–1297
van Noordwijk AJ (1984) Quantitative genetics in natural populations of birds illustrated with examples from the Great Tit, *Parus major.* In: Wöhrmann K, Loeschcke V (eds) Population biology and evolution. Springer, Berlin Heidelberg New York, pp 67–79
van Noordwijk AJ, de Jong G (1986) Acquisition and allocation of resources: their influence on variation in life history tactics. Am Nat 128:137–142

van Noordwijk AJ, van Balen JH, Scharloo W (1981) Genetic variation in egg dimensions in natural populations of the Great Tit. Genetica (The Hague) 55:221–232

Rendel JM (1959) Canalisation of the scute phenotype of *Drosophila*. Evolution 13:425–439

Scharloo W (1962) The influence of selection and temperature on a mutant character (CI^D) in *Drosophila melanogaster*. Arch Neerl Zool 14:431–512

Scheiner SM, Goodnight CJ (1984) The comparison of continuous environmental variation phenotypic plasticity and genetic variation in populations of the grass *Danthonia spicata*. Evolution 38:845–855

Schmalhausen II (1949) Factors of evolution. Blakiston, Philadelphia

Stearns SC (1983) The evolution of life-history traits in mosquitofish since their introduction to Hawaii in 1905: rates of evolution, heritabilites, and developmental plasticity. Am Zool 23:455–463

Stearns SC, Koella JC (1986) The evolution of phenotypic plasticity in life-history traits: prediction of reaction norms for age and size at maturity. Evolution 40:893–913

Thomas VG (1987) Variation in food supply: condition and clutch size, Proc XIX Congr Int Ornith Ottawa 1986 (in press)

Woltereck R (1909) Weitere experimentelle Untersuchungen über Artveränderung, speziell über das Wesen quantitativer Artenunterschiede bei Daphniden. Verh D Zool Ges 1909: 110–172

Chapter 5 Three Approaches to Trade-Offs in Life-History Evolution

M. R. Rose[1,2], P. M. Service[1,3], and E. W. Hutchinson[1,2]

Introduction

The problem of trade-offs between characters is in some way relevant to almost every topic in evolutionary biology. Although we wish to address some of the fundamental issues relating to this problem, evidently we cannot discuss all such points of relevance. We will take the natural way out and concentrate on the kind of questions that we work on in our research. However, we will try to make some effort to do more than advertise our own work.

The particular focus of our work is on life-history evolution. There are two reasons why this work is of general interest for the broader questions of trade-offs in adaptation: (1) life histories are directly related to fitness, giving them a kind of centrality for neo-Darwinian thought; and (2) life-history research has frequently revolved around questions relating to constraints, although under different headings: "reproductive effort", "the cost of reproduction", "antagonistic pleiotropy", and so on. We too will be raising all these issues.

Here, we will structure our discussion around three alternative approaches which have been used to address the problem of evolutionary constraints: (1) optimization theory predicated on some notion of allocation of reproductive effort; (2) mutation-selection balance with a noroptimal selection regime; and (3) antagonistic pleiotropy giving rise to protected polymorphism. We will discuss each of these approaches as distinct theoretical conceptions, in order, and then turn to a discussion of experimental evidence, primarily our own, bearing on how well these alternatives work in an empirical context. Finally, we will propose new experiments to further elucidate the scientific situation.

---

[1] Department of Biology, Dalhousie University, Halifax, Nova Scotia, Canada B3H 4J1
[2] Present Address: School of Biological Sciences, University of California, Irvine, California 92717, USA
[3] Present Address: Department of Genetics, University of California, Davis, California, 95616, USA

Genetic Constraints on Adaptive Evolution
Ed. by V. Loeschcke
© Springer-Verlag Berlin Heidelberg 1987

Optimization Theory

There are few bodies of mathematics that seem to give applied mathematicians more day-to-day pleasure than the calculus of variations and its various congeners (linear programming, stochastic programming, and so on). There are probably many reasons for this. The problems of optimization can be well-defined, mathematically. There have been many years of work of this kind in engineering, so that there are a fair number of tricks of the trade lying around for use.

It was only natural that once Gadgil and Bossert (1970) first put the problem of life-history evolution into the context of optimal allocation there would be a flurry of theoretical papers playing theme and variations on their basic ideas (e.g., Schaffer 1974; Taylor et al. 1974; Leon 1976; review in Charlesworth 1980, pp 231–252), a fashion which has continued into the present day (e.g., Taylor and Williams 1983).

This research has raised a critical question for scientists, a question that applied mathematicians can of course ignore: what is the empirical significance of optimization theory for life-history evolution? This is an issue spiritedly raised in the classic "Spandrels of San Marco" paper by Gould and Lewontin (1979). A more balanced treatment is to be found in Maynard Smith's (1978) review paper on the topic. The obvious problem is, how can optimization theories of evolution be considered scientific if there is no pattern in the data which can't be accounted for on the basis of some optimization model, providing only that additional constraints are introduced? If this is the case, then optimization theory has the irrefutable qualities of astrology, something which Popper (e.g. 1963) has soundly denounced as non-scientific. Maynard Smith has in a way finessed this problem. He concludes that we cannot test the fundamental assumption of ecological applications of optimization theory, that evolution optimizes, we can only test for the particular constraint hypotheses needed to create a successful explanation of observed characters in terms of the expected results of evolutionary optimization.

It seems to us that, for research on life-history evolution, this methodological resolution is valid only within the limited empirical universe of early 1970's research (vid. Stearns 1976, 1977), but that the view of evolution implicit in this empirical universe is pre-Mendelian. One is on a barren Darwinian plane where the only forces that resist the achievement of perfect life histories by selection are "trade-offs", where these trade-offs are not defined in terms of Mendelian hypotheses. This suggests that optimization theory is not just tautological, it is also mechanistically empty. None, or at least few, of us dispute that evolution proceeds by the combined effects of selection, drift, and mutation (where these may take place within genomes as well as within populations) acting on Mendelian variation. Optimization theory does not really have a sound derivation from axioms of this kind. There are contexts where it has some mechanistically-explicit referents. One of these is an asexual population of infinite size which before selection has an infinite supply of genetic variants, covering the relevant phenotype space, no new mutations occurring once selection has begun. However, since the theory is routinely applied to vertebrate life histories, it may be doubted that such points of reference are really adequate to justify the use of optimization theory.

Another way to show the supersession of a body of theory is to demonstrate the existence of questions directly raised by extant research concerning which the theory in question simply says nothing. This is related to the question of the "power" or "positive heuristic" (cf. Lakatos 1970) of a theory. Our point is not just that life-history optimization theory tends to be vacuous and spurious, but that it is rapidly becoming irrelevant. A great deal of empirical life-history research now concerns genetic mechanisms underlying life-history variation, limits to selection response, and so on, as will be discussed below. All of these issues are of importance for a neo-Darwinian characterization of life-history evolution, yet they do not normally surface in optimization theory or the empirical tests of it.

In any event, by this point we hope that we have adequately conveyed our view of optimization theory applied to life histories, a view which its practitioners will undoubtedly find uncharitable.

Mutation-Selection Balance with Noroptimal Selection

One of the most natural ways of approaching the problem of constraints within the context of population genetics theory is the noroptimal selection-mutation balance theory first developed by Kimura (1965), with subsequent elaboration and refinements due to Lande (e.g. 1977), Fleming (1979), and Turelli (1984), among others (e.g. Nagylaki 1984). Unlike optimization theory, this body of theory has been explicitly worked out in population-genetic terms, although many refinements undoubtedly await (cf. Nagylaki 1984). The central assumptions are as follows: the absence of dominance and epistasis; stochastic independence of purely environmental variation with respect to genetic variation, with a Gaussian distribution of effects, and variance V_E; viability selection; sufficiently many alternative allelic states at any contributing locus, such that mutation can be treated as a perturbation on a continuous phenotypic scale; and noroptimal selection such that, if we let y be the phenotypic value on a scale translated such that y = 0 is the optimal value, fitness is given by

$$w(y) \quad = \quad \exp[-y^2/2V_s],$$

were V_s is a scaling parameter giving the intensity of selection.

There is some controversy as to the implications of this work for patterns of genetic variation in natural populations (e.g. Turelli 1984). In the hands of Lande (e.g. 1976), it has been used to argue that high levels of additive genetic variability (vid. Barker & Thomas, this vol.), and thus heritabilities around 0.5, could be due to mutation-selection. By contrast, Turelli (1984, p. 188) has argued that with realistic parameter values and better approximations, the kind of heritabilities which could be maintained by mutation with noroptimal selection are on the order of 0.10 or less.

Antagonistic Pleiotropy

A third body of theory lacks the mathematical sweep and glamour of the two approaches just sketched. One of the simplest ways to model the possibility of evolutionary trade-offs is to suppose allelic effects of opposed direction on distinct fitness components. Consider two alleles segregating at a locus, where there are just two fitness components which add together to determine fitness itself. The two alleles have pleiotropic effects on both fitness components, but these are opposed in direction, giving rise to a selective "antagonism". Without loss of generality, this situation can be represented as follows:

Genotypes:	$A_1 A_1$	$A_1 A_2$	$A_2 A_2$
Fitness components:			
W_1	$1 + h_1 a$	1	$1 - a$
W_2	$1 - b$	1	$1 + h_2 b$
Fitness:			
$W_1 W_2$	$1 - b + h_1 a - h_1 ab$	1	$1 - a + h_2 b - h_2 ab$

Overdominance, and thus genetic polymorphism at selective equilibrium, results if the h_i parameters are sufficiently small. This means that there must be some directional dominance giving rise to a pattern of recessive deleterious gene effects. At selective equilibrium the additive genetic correlations (r_A's) will equal -1 (Rose 1982). Though models of this type require some degree of directional dominance in order to achieve genetic polymorphism at selective equilibrium (Rose 1982, 1983b, 1985), and thus high heritabilities and negative genetic correlations between fitness components, the resulting levels of nonadditive genetic variance at selective equilibrium need not be large in magnitude (Rose 1982, 1985).

This theory has been generalized over a range of well-understood population-genetics models: arbitarily large numbers of fitness components that are additive or multiplicative; one triallelic locus; two diallelic loci with epistasis; sex-specific effects; and overlapping generations (Rose 1982, 1983b, 1985). This work shows that "trade-offs" which give rise to antagonistic pleiotropy can in turn lead to the maintenance of abundant additive genetic variability for fitness-related characters, in association with negative genetic correlations between these characters. The empirical requirements of this theory are therefore quite clear; abundant additive genetic variability should be associated with negative genetic correlations between the genetically variable characters.

A Stroll in the Jungle

Scientists have widely varying opinions about the relationship between theory and data. Let us render our view in the following simile. For us, theories are like sheltered, upper-class, Victorian schoolgirls, dressed up entirely in white and severely scrubbed. (This reflects our impression of their cleanliness and lack of suitability for the dirty business of life itself.) Confronting theories with data is like taking these schoolgirls on an expedition to a tropical rainforest, with careful chaperoning by the proponents of these theories. The chaperones will of course make every attempt to avoid exposing the girls to any of the hazards of the jungle, but the occasional mirthful accident in which white linen becomes hopelessly muddied is inevitable. Of greater concern are those incidents where some ravenous beast comes from the shadows and carries off one of these little ladies to be devoured.

Being Darwinists, our inclinations are to foster selection acting so as to increase the frequency of the fittest, even if it means the destruction of one of the charming creations of the theoreticians. This is an ugly process, but we feel that it is necessary nonetheless. Here we will take the three theories outlined above on a brief stroll through the jungle, giving them opportunities to display the extent to which they are, or are not, suited to survive in the cruel and violent world of evolutionary reality. To give a foretaste of what will happen on this stroll, let us say at the outset that none of our dainty girls is going to escape without scratches and bruises, and our view is that one of them has been carried off into the bushes.

The theory at greatest risk, in our view, is optimization theory. The threat to it is at once obvious: if evolution has optimized life histories with any precision, why is there so much additive genetic variation for life-history characters segregating in outbred populations? The evidence for this empirical point has been summarized before (e.g., Rose 1983a), and we will not go over it again here. Chaperones of this theory can try to protect her by claiming that changing and/or spatially varying environments could explain these results, in that such environments can maintain genetic variation because the optimization process of evolution never proceeds to the appropriate equilibrium, since the latter is always moving (vid. Felsenstein 1976). There are two lines of evidence against this line of defense. The first is that long-established laboratory populations can maintain high equilibrium levels of genetic variation in life-history characters (e.g., Rose and Charlesworth 1981a; Rose 1984a; Service and Rose 1985). The second is that the required pattern of dependence of quantitative genetic variability on environmental variation has been tested for and found lacking (e.g., Riddle et al. 1986). All evidence of this kind suggests that this particular body of theory is in mortal danger.

A different problem is that it appears to be unable to cope with the strange and difficult geography of empirical work on life-history evolution. What does optimization theory have to say to us about the prospects for success when we select on particular life-history characters, or their determinants? What does it suggest about the limits to such selection? What novel experimental approaches does it suggest, beyond seeing if its predictions "fit"? It seems to us that this little "girl" is a naive waif hardly suited to last in the jungle that we are proceeding through scientifically.

The remaining two approaches cannot be fed to the tigers quite so easily. Indeed, one of the salient features of the mutation-selection balance theory is that it hinges on the magnitudes of two parameters which are notoriously difficult to measure: mutation rates and selection intensities. [This same empirical difficulty largely vitiated the neutralist-selectionist debate before modern DNA sequence data became available (Lewontin 1974).] At present, we know of only two pieces of evidence which can scratch the bare legs of this theory. Rose and Charlesworth (1981a) studied the age-specific additive genetic variance of female fecundity in *Drosophila melanogaster* from an early age to late in life. The importance of this work lies in the fact that, if one imagines a process of mutation-selection balance affecting the evolution of age-specific fecundity, in the absence of pleiotropy between early and late ages, then one is studying the dependence of the equilibrium variance of a fitness-related character along a gradient on which the intensity of natural selection must fall steeply (cf. Hamilton 1966; Charlesworth 1980). Assuming that mutation rates of genes with age-specific effects on fecundity do not coincidentally fall too along this same gradient, which a desperate chaperone might insist upon, then there should be significant increase in the additive genetic variance with age. As shown in Fig. 1, there was no such increase. This suggests that some other process was responsible for the maintenance of the abundant additive genetic variance for this character.

A second problem for this theory, particularly in its guise as a foundation for further theory assuming maintenance of heritable genetic variation in the face of selection, is that of limits to selection. The normal assumption of a breeder, plant or animal, when a stock quickly ceases to respond to selection is that the organisms were originally fairly inbred to begin with. For the *D. melanogaster* population studied in Rose (1984a), this is not a reasonable assumption, given what had already been found of its supply of genetic variation affecting life-history characters. Yet a

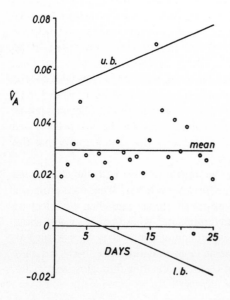

Fig. 1. Pattern of additive genetic variance for daily fecundity of adult *Drosophila melanogaster* as a function of age at which fecundity was assayed. (The first day of assay was a few days after pupal emergence.) Results are from an experiment reported in Rose and Charlesworth (1981a)

plateau was reached within five to eight generations of artificial selection for increased early fecundity (Rose 1984a). This result suggests that Lande's (1976) basic interpretation of the sources of genetic variation for characters of this kind is in need of revision. Other independent evidence (Rose and Charlesworth 1981a, b; Rose 1984b; Service and Rose 1985) indicated that abundant additive genetic variation was being maintained for early fecundity by antagonistic pleiotropy, and others have obtained similar results (e.g. Luckinbill et al. 1984). To be specific, genetically enhanced fecundity appears to exact a cost in resources required for survival, possibly lipids specifically (Service et al. 1985; Service and Rose 1985). If there are loci subject to antagonistic pleiotropy responsible for the additive genetic variance of a specific character, then artificial selection upon that character may be opposed by natural selection, giving rise to difficulties with sustained response to selection.

There are a number of points on which the antagonistic pleiotropy theory can be attacked. The most important of these is that it is often found that the genetic correlations between fitness characters are positive, or approximately zero, rather than negative (e.g. Giesel 1979; Gisel and Zettler 1980; Giesel et al. 1982; Murphy et al. 1983; Bell 1984a, b). We have been chaperoning the theory on this point, particularly by demonstrating that inbreeding depression and genotype-environment interaction could have given rise to these results artifactually (Rose 1984c; Service and Rose 1985; Rose and Service 1985). However, these are patently defenses of a maiden in whose virtue we are interested; therefore, the possibility must be admitted that some of these findings against antagonistic pleiotropy may be "real", the ambiguities in their interpretation notwithstanding.

Another blow against this theory is that there is little evidence for the pattern of dominance required if antagonistic pleiotropy is to be used as an explanation for abundant additive genetic variation in fitness components (Rose 1982, 1983b, 1985). Clare and Luckinbill (1985) have looked for it in the inheritance of longevity in *Drosophila melanogaster* and not found it. We have looked for dominance in *D. melanogaster* fitness characters too, in the inheritance of longevity, early fecundity, resistance to starvation, and weight of ovaries. So far, all our results are negative; dominance of such a type as to facilitate the maintenance of genetic polymorphism via antagonistic pleiotropy has not been found (Hutchinson and Rose in prep.). These results are not strictly fatal; it is possible to meet the conditions required for protected polymorphism with additivity and antagonistic pleiotropy (Rose 1982, 1983b). But it is difficult to do so at many loci simultaneously, suggesting limits to the role of antagonistic pleiotropy in maintaining genetic variation for fitness-related characters.

A "Crucial" Experiment

Many empirical scientists aspire to perform "crucial" experiments, experiments which will falsify one theory while corroborating another. The practical situation, however, is that there is always room for chaperones of theories to rush about in order to save their theories (Popper 1959; Lakatos 1970). The most that crucial experiments can do is make the efforts of chaperones obviously ludicrous. Some "theory-saving" post hoc hypothesis can always be invented, but the greater the peculicarity of such hypotheses, the worse the "saved" theory looks (Lakatos 1970).

We ourselves set about contriving an experiment which would put our own theoretical approach, antagonistic pleiotropy, at risk of falsification. We also feel that this experiment bears on the validity of the other two alternative approaches. While the results can, of course, be explained away using additional ad hoc hypotheses, we feel that what we have found at least begins to make life rather difficult for single-minded devotees of any one of these bodies of theory.

The nature of the experimental material and methods should be described first. We have been using naturally selected lines of *D. melanogaster* to dissect out the subsidiary characters responsible for enhancing adult survival (Service et al. 1985). These lines fall into two types, which we call "B" and "O", there being five lines of each type, distinguished by subscripts. These lines were created by the imposition of two different regimes of culture maintenance in the laboratory, starting from the same generation of a common base population. Since then, the B lines have undergone 14-day discrete generations, at $25^{\circ}C$ with abundant food. The O lines have had the length of their discrete generations progressively increased from 14 to 70 days, at $25^{\circ}C$ with abundant food, by rapid turnover of medium in order to prevent eggs laid before 70 days of age from contributing to the next generation. This culture regime has led to the evolution of postponed senescence (Rose 1984b), as exhibited in increased mean longevities of sampled flies, in association with reduced early fecundity, in conformity with earlier results from these same flies (Rose and Charlesworth 1981a, b) and the results of others working independently (Luckinbill et al. 1984).

We wanted to dissect the evolutionary physiological genetics of this postponed senescence, and its associated antagonistic pleiotropy, in more detail. Accordingly, we set about looking for subsidiary physiological characters which had evolved in parallel with postponed senescence. Our initial focus was on increased resistance to short-term lethal stresses at a variety of adult ages. We found consistent B-O differences with respect to resistance to starvation, resistance to desiccation, and resistance to low levels of ambient ethanol, the first two being consistent over a range of adult ages, the last tested only at early ages (Service et al. 1985).

In January 1984, we derived a series of five replicated reversed selection lines, called R_1 to R_5, from O_1 to O_5, respectively. We reinstated selection for early fertility in these R lines, relaxing selection for enhanced adult survival. Four characters were examined in the young adults of every other generation for 22 generations: 24 h fecundity, time to death due to starvation, time to death to desiccation, and time to death due to exposure to 15% ambient ethanol. After the twenty-second generation, we continued to assay starvation resistance differences, although our procedures changed somewhat, so we will not report the data here.

We analyzed the data by pairing each of the five R lines with its corresponding B control, and considering both the difference between the means of each R and B pair and the ratio of the means of the R and B lines. The results for the ratio data are summarized in Fig. 2, where the means of the five ratios for each character are shown as a function of generation. For statistical tests, we computed the regression coefficient of the R/B ratio for each population pair on generation. The regression slopes for each character were then used as raw data for t-tests of their significance. Thus, in our analysis the evolution of a single character in one R line over 22 generations was taken as a single datum. Our significance threshold was set at a probability of 0.05. Early fecundity had been depressed as a result of selection for enhanced adult survival (Rose 1984b), but it increased significantly once selection for early reproduction was reintroduced, as shown in Fig. 2. At the same time, starvation resistance fell significantly, as we expected from previous estimates of a substantial negative genetic correlation between the two characters (Service and Rose 1985). Ethanol and desiccation resistance did not exhibit any statistically significant change over this period. Moreover, starvation resistance has not returned to the level of the controls, even though fecundity has, and our later data (not shown) indicate that it will not do so soon. Therefore, this experiment suggests that antagonistic pleiotropy between early fertility and adult survival did not play a role in determining the evolution of ethanol and desiccation resistance, while its role in shaping the evolution of starvation resistance was only partial. While antagonistic pleiotropy might ultimately provide an adequate explanation for the genetic variation for fecundity, it clearly will not do as a general explanation for these results.

The results for fecundity are in accord with our previous interpretations (Rose and Charlesworth 1981a, b; Rose 1984b; Service and Rose 1985). There appears to be antagonistic pleiotropy between early fecundity and adult survival. The starvation resistance results broadly fit this pattern, in that the character declines as fecundity increases. A problem with these results, however, is that starvation resistance in the R's has not returned to the levels of the B controls, unlike the return to control levels exhibited by fecundity. What is worst of all for the antagonistic pleiotropy theory is that ethanol and desiccation resistance have not changed significantly upon relaxation of selection at later ages and reintroduction of culture selection for enhanced early fertility. This suggests that *none* of the genes involved in enhancing these characters, as a result of selection for increased adult survival rates, exhibit antagonistic pleiotropy with respect to early fitness components, at least in our experimental system.

Like good chaperoning scientists, we can think of reasons why these results were obtained which would save our theory. The obvious one is to suppose that there has been fixation of the alleles fostering ethanol and desiccation resistance in the original O lines, so that reversed selection could not change their frequency, at least until appropriate mutations have occurred. Perhaps the same idea could be used to explain the failure of starvation resistance to return to the control level as well. But we do not find this particularly convincing. There are probably multiple loci

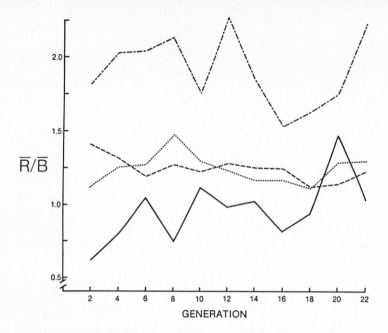

Fig. 2. Results of reversed selection on adult life history in five derivatives (R_1 to R_5) of stocks exhibiting enhanced stress resistance. The data plotted are the averages of the five ratios of R_i to B_i, its paired control, for each of four characters: early 24 h fecundity (*solid line*); starvation resistance (*dashed line*), desiccation resistance (*dotted line*), and ethanol resistance (*dasheddotted line*). The experimental generation is indicated on the x-axis. See the text for explanation of experimental procedures

involved in the differentiation of O and B stress-resistance characters, the likely effect of selection not being fixation of all selectively favored alleles in our large O populations.

The Present Situation

Our present view is that *both* antagonistic pleiotropy and mutation-selection balance play some role in the evolution of life histories. Since life histories are central, we therefore conclude that both population genetic mechanisms act to constrain evolution, generally.

A number of qualifications and elaborations need to be appended to this bald conclusion. Our view is that our work amounts to a good, although not absolutely definitive, demonstration of antagonistic pleiotropy as a potential evolutionary force in *Drosophila* laboratory populations. Others might question whether or not this is in fact a representative case. We would point to the meadow grass work of Law (Law et al. 1977; Law 1979) for evidence that such antagonisms are of more general importance. We would also emphasize that our work is not the only

Drosophila research which gives results of this kind. Luckinbill et al. (1984) have closely parallel findings, while Simmons et al. (1980) used quite different methods to come to comparable conclusions. It could also be said that the classic sickle-cell anemia polymorphism in sub-Saharan Africa is an instance of antagonistic pleiotropy, in that there are two pleiotropic effects of sickling, deficient oxygen transport and malarial resistance, which give rise to the kind of overdominance required by the antagonistic pleiotropy theory. All this notwithstanding, we regard antagonistic pleiotropy as an inadequate basis for the explanation of most quantitative genetic variation. It appears from the work discussed above that the pattern of directional dominance of effect on single fitness components, which is important for achieving protected polymorphism (e.g. Rose 1983b), is not generally achieved. Most importantly, there appear to be some characters which can evolve in ways apparently free of the required antagonistic fitness-component effects, such as the desiccation and ethanol resistance characters of Service et al. (1985).

We also conclude that evidence is beginning to crop up for mutation as a significant source of genetic variation for quantitative characters. There is no other straight-forward interpretation of the reversed selection results that we present here. This is not to say that we have evidence for the sort of noroptimal selection-mutation balance studied by Lande et al. The genetic variation affecting ethanol and desiccation resistance that responded to selection for increased adult survival could have been virtually neutral, in principle. However, we would tend to subscribe to a view like Kimura's (1983, p. 142–143), in which weak stabilizing selection at the phenotypic level is associated with near neutrality of allelic variants at any one locus. Thus, we would expect, over the long run, that mutation-selection balance would return the ethanol and desiccation resistance characters of the R populations discussed above to the levels characteristic of the B populations. This would, of course, be quite different from the rapid return to the B level exhibited by fecundity in Fig. 2. As for the starvation character, our opinion right now is that it could be subject to both evolutionary mechanisms. However, it is obvious that we are only beginning to tackle the problems associated with dissecting the evolutionary quantitative genetics of loci not subject to antagonistic pleiotropy.

Let us briefly return to the subject of optimization theory. In our opinion, the level that mathematical and experimental research has reached in the study of life-history evolution is such that optimization theory is simply outmoded in the study of life-history evolution. It offers little in the way of predictions about evolutionary mechanisms, and still less does it suggest fruitful avenues for further experimentation. From this point on, we expect it to become a mere distraction, at least where research on life-history evolution is concerned. Other topics within evolutionary biology might by illuminated by the pale light that it casts, but life-history evolution is no longer one of those.

What Now?

The evidence that both antagonistic pleiotropy and mutation-selection balance play a role in the maintenance of genetic variability suggests to us that investigators should now seek ways to distinguish between the three possible cases for quantitative genetic variation affecting specific characters: (1) predominance of antagonistic pleiotropy as a force maintaining genetic variation; (2) predominance of mutation-selection balance as a force maintaining genetic variation; and (3) combinations of these two forces, perhaps with other mechanisms too, in the maintenance of genetic variation. Further, it seems to us that our fellow workers should, for the time being, concentrate on cases (1) and (2), since these will be least ambiguous.

Given this imperative, it becomes necessary to discover methods of sorting out patterns of pleiotropy and selection. At this point, it might be helpful if we mention some of the experimental strategies that we have found useful. The most fundamental decision that we have made in our work is to use an outbred laboratory population which has been kept at large census numbers in a stable environment. This frees us from a host of problems which are otherwise difficult to elude (Rose and Service 1985). The first of these is the problem of maintenance of genetic variation as a result of "coarse-grained" temporal or spatial variation. Certainly our laboratory culture system does not afford such variation. Secondly, we are confident of evolutionary equilibrium in our population, an important requirement because the lack of evolutionary equilibrium in a population makes it useless for comparing theoretical alternatives which are readily distinguishable only at evolutionary equilibrium. Thirdly, and partly a corollary of the second point, we do not need to worry about genotype-environment interaction as much as those who sample from the wild (cf. Service and Rose 1985). Fourthly, we do not need to worry overly about inbreeding depression (vid. Rose 1984c). In effect, our system allows us to avoid the most vitiating artifacts plaguing research on fitness characters.

A more recent move of ours has been to get away from "aggregated" characters which are the result of a great many subsidiary characters. Longevity is an obvious example of such a character. Indeed, so long as we confined ourselves to the relationship between longevity and early fertility, it very much appeared as if antagonistic pleiotropy was the whole story (Rose and Charlesworth 1981a, b; Rose 1984b). As shown here, with consideration of an array of subsidiary characters contributing to longevity, we have been able to discover major differences between them with respect to their evolutionary molding. Accordingly, we have concluded that incisive evolutionary analysis requires the use of such underlying characters. However, this is not to say that we think that it makes sense simply to start with such characters and assume that they are relevant to fitness in some way. Our approach has instead been to proceed from quantitative genetic characterization of aggregated life-history characters to their constituent characters by careful probing for consistent correlations (vid. Rose et al. 1984; Service et al. 1985). A great many characters which we initially thought might be important in our system, like total body size and resistance to heat stress, have not turned out to be involved in the segregating genetic variation that we have studied. Therefore, we would suggest an explicitly *hierarchical reductionist* approach, beginning at the top and slowly edging downward. [This may be contrasted with the dogmatic reductionist approach which starts with molecules or cell cultures, in the absence of any clear interpretation of the significance of the studied process(es) in the determination of the quantitative genetic variation of fitness components.] This method is laborious, but we have found that it gives results that we can trust.

When one gets to classically "physiological" characters, the temptation may grow to abandon quantitative or evolutionary genetic methods and set about doing more conventional "straight physiology" or cell biology. We think that this is a mistake. Continuing with genetic principles of analysis remains possible and, we would suggest, highly productive. In particular, it will probably remain the case that special mutant stocks affecting the characters revealed in the course of hierarchical reduction will be no more informative than such mutants have normally been in unraveling the genetic variation affecting fitness itself. The reason for this is simple. It is high-fitness alleles alone which are of interest, and visibly mutant alleles are normally associated with substantial fitness depression. High-fitness alleles cannot be handled the way we might handle *white*, e.g., in *D. melanogaster.*

Finally, we should say that we do not suppose that our methods are the only ones which will shed light on the nature of the mechanisms responsible for constraints in life-history evolution. There are undoubtedly many fine experiments out there waiting to be formulated, experiments which more people should be trying to do.

Summary

Three different theoretical approaches to trade-offs in life-history evolution are described: optimization, noroptimal mutation-selection balance, and antagonistic pleiotropy. It is questioned whether optimization theory is well founded at the population genetic level. The extant empirical evidence is then brought to bear on these theories, with a view to challenging each as much as possible. Optimization models face the problem of accounting for the abundant heritable fitness component variation which appears to be maintained in some populations. Mutation-selection theory is ostensibly refuted, at least as a completely general theory, by the absence of any relationship between the intensity of selection and age-specific additive genetic variation in *Drosophila melanogaster;* a substantial fall in the intensity of selection does not appear to lead to an increased level of equilibrium additive genetic variance. In addition, there is a plethora of evidence supporting the antagonistic pleiotropy theory, which therefore by inversion threatens the mutation-selection balance model. The antagonistic pleiotropy theory is confronted by a wide range of ostensibly falsifying results with respect to its prediction of negative genetic correlations between life-history characters. The present chapter also reports a new experiment, involving correlated responses to natural selection on early reproduction, where the characters concerned foster adult survival. The results of this experiment indicate that *both* antagonistic pleiotropy and mutation-selection balance may play a considerable role in the evolution of life-history characters, and fitness components, generally.

Acknowledgments. We thank J.S.F. Barker, A. Clark, V. Loeschke, and M. Turelli for their many comments on an earlier draft. We are grateful to L.E. Johnston, D.M. Lane, and A.J. Shaw for many hours of technical support. This research was supported by an operating grant from NSERC of Canada, while the first author was an NSERC University Research Fellow.

References

Bell G (1984a) Measuring the cost of reproduction. I. The correlation structure of the life table of a plankton rotifer. Evolution 38:300–313

Bell G (1984b) Measuring the cost of reproduction. II. The correlation structure of the life tables of five freshwater invertebrates. Evolution 38:314–326

Charlesworth B (1980) *Evolution in age-structured populations.* Cambridge University Press, London

Clare MJ, Luckinbill S (1985) The effects of gene-environment interaction on the expression of longevity. Heredity 55:19–29

Felsenstein J (1976) The theoretical population genetics of variable selection and migration. Annu Rev Genet 10:253–280

Fleming WH (1979) Equilibrium distributions of continuous polygenic traits. SIAM J Appl Math 36:148–168

Gadgil M, Bossert WH (1970) Life historical consequences of natural selection. Am Nat 102: 52–64

Giesel JT (1979) Genetic co-variation of survivorship and other fitness indices. Exp Gerontol 14:323–328

Giesel JT, Zettler EE (1980) Genetic correlations of life historical parameters and certain fitness indices in *Drosphila melanogaster:* r_m, r_s, diet breadth. Oecologia 47:299–302

Giesel JT, Murphy PA, Manlove MN (1982) The influence of temperature on genetic interrelationships of life history traits in a population of *Drosophila melanogaster:* what tangled data sets we weave. Am Nat 119:464–479

Gould SJ, Lewontin RC (1979) The spandrels of San Marco and the Panglossian paradigm: a critique of the adaptationist programme. Proc R Soc B 205:581–598

Hamilton WD (1966) The moulding of senescence by natural selection. J Theor Biol 12:12–45

Kimura M (1965) A stochastic model concerning the maintenance of genetic variability in quantitative characters. Proc Natl Acad Sci USA 54:731–736

Kimura M (1983) *The neutral theory of molecular evolution:* Cambridge University Press, London

Lakatos I (1970) Falsification and the methodology of scientific research programmes. In: Lakatos I, Musgrave A (eds) *Criticism and the growth of knowledge.* Cambridge University Press, London

Lande R (1976) The maintenance of genetic variability by mutation in a polygeneic character with linked loci. Genet Res 26:221–235

Lande R (1977) The influence of the mating system on the maintenance of genetic variability in polygenic characters. Genetics 86:485–498

Law R (1979) The cost of reproduction in annual meadow grass. Am Nat 113:3–16

Law R, Bradshaw AD, Putwain PD (1977) Life-history variation in *Poa annua.* Evolution 31: 233–246

Leon JA (1976) Life histories as adaptive strategies. J Theor Biol 60:301–306

Lewontin RC (1974) *The genetic basis of evolutionary change.* Columbia University Press, New York

Luckinbill LS, Arking R, Clare MJ, Cirocco WC, Buck SA (1984) Selection for delayed senescence in *Drosophila melanogaster*. Evolution 38:996–1003

Maynard Smith J (1978) Optimization theory in evolution. Annu Rev Ecol Syst 9:31–56

Murphy PA, Giesel JT, Manlove MN (1983) Temperature effects on life history variation in *Drosophila simulans*. Evolution 37:1181–1192

Nagylaki T (1984). Selection on a quantitative character. Chakravarti A (ed) In: *Human population genetics: the Pittsburgh symposium*. Van Nostrand Reinhold, New York, pp 275–306

Popper KR (1959) *The logic of scientific discovery*. Hutchinson Ross, London

Popper KR (1963) *Conjectures and refutations*. Harper & Row, London

Riddle RA, Dawon PS, Zirkle DF (1986) An experimental test of the relationship between genetic variation and environmental variation in Tribolium flour beetles. Genetics 113: 391–404

Rose MR (1982) Antagonistic pleiotropy, dominance, and genetic variation. Heredity 48:63–78

Rose MR (1983a) Theories of life-history evolution. Am Zool 23:15–23

Rose MR (1983b) Further models of selection with antagonistic pleiotropy. Freedman HI, Strobeck C (eds) In: *Population biology*. Springer, Berlin Heidelberg New York, pp 47–53

Rose MR (1984a) Artificial selection on a fitness-component in *Drosophila melanogaster*. Evolution 38:516–526

Rose MR (1984b) Laboratory evolution of postponed senescence in *Drosophila melanogaster*. Evolution 38:1004–1010

Rose MR (1984c) Genetic covariation in *Drosophila* life history: untangling the data. Am Nat 123:565–569

Rose MR (1985) Life history evolution with antagonistic pleiotropy and overlapping generations. Theor Popul Biol 28:342–358

Rose MR, Charlesworth B (1981a) Genetics of life-history in *Drosophila melanogaster*. I. Sib analysis of adult females. Genetics 97:173–186

Rose MR, Charlesworth B (1981b) Genetics of life-history in *Drosophila melanogaster*. II. Exploratory selection experiments. Genetics 97:187–196

Rose MR, Service PM (1985) Evolution of aging. Rev Biol Res Aging 2:85–98

Rose MR, Dorey ML, Coyle AM, Serice PM (1984) The morphology of postponed senescence in *Drosophila melanogaster*. Can J Zool 62:1576–1580

Schaffer WM (1974) Selection for optimal life histories: the effects of age structure. Ecology 55: 291–303

Service PM, Rose MR (1985) Genetic covariation among life-history components: the effect of novel environments. Evolution 39:943–945

Service PM, Hutchinson EW, MacKinley MD, Rose MR (1985) Resistance to environmental stress in *Drosophila melanogaster* selected for postponed senescence. Physiol Zool 58: 380–389

Simmons MJ, Preston CR, Engels WR (1980) Pleiotropic effects on fitness of mutations affecting viability in *Drosophila melanogaster*. Genetics 94:467–475

Stearns SC (1976) Life history tactics: a review of the ideas. Q Rev Biol 51:3–47

Stearns SC (1977) The evolution of life history traits: a critique of the theory and a review of the data. Annu Rev Ecol Syst 8:145–171

Taylor HM, Gourley RS, Lawrence CE (1974) Natural selection of life history attributes: an analytical approach. Theor Popul Biol 5:104–122

Taylor PD, Williams GC (1983) A geometric model for optimal life history. In: Freedman HI, Strobeck C (eds) *Population biology*. Springer, Berlin Heidelberg New York

Turelli M (1984) Heritable genetic variation via mutation-selection balance: Lerch's zeta meets the abdominal bristle. Theor Popul Biol 25:138–193

Chapter 6 Pleiotropy in Dynamical Parameters of Models on the Evolution of Simple Phenotypes

F. B. CHRISTIANSEN [1]

Introduction

The Darwinian theory of anagenetic evolution is based on the interplay of two phenomena: the existence of hereditary phenotypic variation in natural populations and the occurrence of natural selection whenever phenotypic variation exists. Constraints on the evolutionary process may therefore be understood as limitations on either of these phenomena. Limitations due to the nature of the hereditary variation is the subject of other contributions in this symposium. Consideration of the nature of the hereditary variation is important in the discussion of natural selection at various levels, but the large-scale limitations imposed by the circumstances of the variation is probably best illustrated by considering the concept of pre-adaptation. Suppose in a given species that natural selection distinguishes among the variations on the typical phenotype in relation to an organ with a given function. The distinction in general is based on the quality of the function of this organ. If the organ changes as evolution proceeds, variations in the now typical phenotype may open up for an entirely different function of the organ and thereby change the nature of natural selection on the variations of the organ. If the new function is favored by natural selection, then in hindsight the organ is said to have been "pre-adapted" for the new function.

This emphasizes that natural selection works on the available variation according to the circumstances of the species and its environment. Theoretical considerations of such phenomena as pre-adaptation, which may lead to what often is described as macro-evolutionary change, are difficult beyond the ascertainment that they are indeed theoretically possible. However, the process of natural selection is inherently the same before, during, and after such an important evolutionary event, because natural selection is a necessary consequence of the existence of phenotypic variation in characters related to the environmental adaptation of the organism. Therefore, we may as well discuss the limitations in an evolutionary stable situation, that is, where simple anagenetic changes occur in a well-defined, simple character with a simple genetic determination of the hereditary component of the variation.

The evolutionary important consequence of natural selection is that the genetic composition, and hence the typical phenotype, of the population changes. Therefore,

[1] Department of Ecology and Genetics, University of Aarhus, Ny Munkegade, DK-8000 Aarhus C, Denmark

Genetic Constraints on Adaptive Evolution
Ed. by V. Loeschcke
© Springer-Verlag Berlin Heidelberg 1987

evolution by natural selection often is studied theoretically by considering the selection induced on the genotypic variation as a result of the connection between the genotype and the phenotype. This is the typical approach of population genetic theory, and it has proven successful in evaluating the immediate genetic constraints on the process of evolution. However, when discussing the ecological constraints on natural selection, the original definition of natural selection as acting on the phenotypic variation in a population should be maintained, as ecology in the broad sense is concerned with the interplay between the individual and its environment. Viewed in this way, the ecological description of an organism in its environment forms the basis for the description of the process of natural selection as necessitated by individual variation.

The description of natural selection is often made in terms of the related variation in the theoretical character "fitness". This is, in the sense of Darwin (1872), the character that is supposed to summarize what is good and what is bad in the individual. Already Darwin realized that fitness is not a simple character. It is composed of the two primary fitness components related to the probability of survival of the individual to sexual maturity and to the fecundity of sexually mature individuals. However, the fecundity of sexually mature individuals is not only determined by the potential to produce offspring, since the probability of becoming a parent may depend on intricate interactions with other individuals of the species giving rise to the process of sexual selection. These multiple facets of the process of natural selection are further complicated by the nature of the indirect Mendelian inheritance in diploid organisms with sexual reproduction, in that the induced selection on the genetic composition is inherently different for the various selection components. Thus, even though fitness is a very intuitively appealing concept, which is valuable in many biological considerations, it should be kept in mind that any evaluation of fitness is an incomplete summary of the process of selection in a natural population (Christiansen 1984b).

For theoretical considerations the dynamical models of population ecology can form the basis of models of natural selection. The transformation of these models into models for the study of evolution has taken different routes. One, which we may term the population-genetics way, models the induced selection on a genotypic variation directly, that is, genotypic variation in the model parameters is considered. This approach is important in the thriving studies of the evolutionary consequences of density- and frequency-dependent selection. On the other hand, we can instead maintain the primary connection between phenotype and fitness as belonging to the subject of ecology and the connection between the genotype and the phenotype as belonging to the subject of genetics, and build models of the evolution of ecologically interesting phenotypes as models within, rather than between, the two subjects. An important difference in the two approaches is the way in which genotypic effects on the phenotype are modeled. The first approach focuses on genetic variation with pleiotropic effects on the ecologically important dynamical parameters. Here, it is customary to judge the generality of the model from the freedom in variation allowed among these pleiotropic effects, so the framework of the model is primarily a genetic one. The second approach focuses on the phenotypic variation on a given character and dynamical pleiotropy of genotypic variation is modeled from the influence of this character on the adaptation of the individual to its environment. Thus, the framework of the model is here primarily ecological, and the generality of the model should be judged in terms of the relations between the phenotype and the environment. This will necessarily restrict the freedom in the variation of the effects of dynamical pleiotropy, as the pleiotropy of the considered genetic variation is deduced from the ecological model. In the end, similar theoretical models may, of course, be reached along the two routes. The basic difference between the two approaches is in the constraints they suggest concerning the necessary simplifications that allow theoretical results to be developed. Therefore, when discussing ecological constraints on the evolutionary process, the second route seems the most natural way to proceed.

This modeling approach may be illustrated by considering some simple arguments concerning the evolution of reproductive parameters in relation to simple phenotypes as the size of an organism at various life stages. The focus of the arguments is the modelling approach, so genetic and ecological complications will be avoided. In relations to the thriving theory of life-history evolution the models therefore may seem unnecessarily simplified, both as models of population genetics and as models of population ecology.

Models of the Growth in Body Size and the Population Dynamics

Consider for simplicity in the arguments a sexual organism with nonoverlapping generations. Suppose further that it breeds by random mating at a discrete time terminating each generation. This assumption produces models of maximum genetic simplicity as everybody breeds at the same time. However, for ecological considerations, of course, the organism has a continued development during each generation cycle from zygotes (formed into eggs or seeds) and on to mature individuals which eventually mate and reproduce.

During development the size of the individual increases. Let $\sigma(t)$ describe the size of an individual at age t, and let the growth of individual size be described in terms of the specific growth rate g, so the equation:

$$\frac{d\sigma}{dt} = \sigma g(\sigma), \tag{1}$$

predicts $\sigma(t)$ as a function of the egg size $s = \sigma(0)$. The adult size is then given as $S = \sigma(T)$, where T is the length of the growth period. We refer to T as the development time of eggs to the adult stage and to S as the initial size of the adult stage. This is a very simple growth model, in that the specific growth rate only depends on the size and not, e.g., on the age of the organism. The time to reach sexual maturity, T, may be a constant independent of the size of the developing individual, but simple size influence on the time to reach sexual maturity is introduced by making T dependent on s.

As an example of the growth model of Eq. (1) we may use the simplest sigmoid growth curve, where the specific growth rate decreases linearly with σ:

$$g(\sigma) \quad = \alpha(1 - \beta\sigma). \tag{2}$$

This model, with a given development time, T, can accomodate a range of growth profiles ranging from indeterminate exponential growth (β small) to very determinate growth with a very weak dependence of the adult size on s (T large compared to $-\log \beta/\alpha$).

The population of developing individuals is decimated due to the death of individuals from various causes. Let x be the number of eggs laid in the population in a given generation, and let u(t) be the number of individuals in the population at age t so u(0) = x. We assume the simple model for density-dependent death of Poulsen (1979), in which the developing organisms die off with a per capita death rate that increases linearly with population density:

$$\frac{1}{u(t)} \quad \frac{du}{dt} \quad = -[d + fu(t)]. \tag{3}$$

The number of adults in this generation is determined as the number of individuals, u(T), in the population by the end of the developmental period. If each of these adults survive with the probability D_A from the end of the developmental period to the time of breeding, and if each breeding individual produces B offspring, then the number of eggs starting the next generation is $x' = BD_A u(T)$.

With these simple assumptions we may produce a variety of models of the evolution of the size of the organism by specifying the fitness parameters d (the density-independent death rate), f (the density-dependent death rate coefficient), and B (the fecundity) as functions of the phenotype s. The ecological constraints are then expressed in terms of the functional relationship between s and the fitness parameters. If density is low, then the survival probability, D, of an individual from egg to the adult stage is:

$$D(s) = \exp\left[-\int_0^T d(\sigma) \ dt\right], \tag{4}$$

so the expected number of offspring of an egg is $B(s)D_A D(s)$. This expected number of offspring is the intrinsic fator for the growth in population size, in that the population grows at low densities when $B(s)D_A D(s) > 1$ and the population dies out when $B(s)D_A D(s) < 1$. It is tempting to conclude that density-independent fitness is simply described by this intrinsic factor, $B(s)D_A D(s)$. In a sexual organism the probability of survival indeed may be considered as an individual phenotype, but the fecundity is necessarily a property of a female-male pair of individuals, so it cannot be treated as an individual phenotype. Nevertheless, for the kind of crude arguments that will be used here, this complication can be neglected, e.g., the later results are precise if fecundity is considered as a female attribute only (Feldman et al. 1983). However, the pair attribute of fecundity should be reintroduced if more detailed models are of concern.

From population genetics we know that a new rare allele introduced at a locus will increase with time if the average $B(s)D_A D(s)$ of the allele in the population is larger than the average $B(s)D_A D(s)$ of the original population. This may seem to be an unnecessarily weak statement as compared to the fundamental theorem of natural selection, which states that the average fitness of the population always increases. However, the fundamental theorem of natural selection holds only for variation in density-independent survival, not, e.g., for variation in fecundity. This weaker result for the initial increase of a rare allele in the density-independent situation extends to the density-dependent case, where the fitness $B(s)D_A D(s)$ is modified by a density-dependent death rate factor,

$$F(s,x) = \exp[-\int_0^T u(t)f(\sigma)\, dt], \qquad (5)$$

to $B(s)D_A D(s)F(s,x)$ in a population where the egg density, typically at the equilibrium density, is given by x. A new allele will then increase in frequency if its average $B(s)D_A D(s)F(s,x)$ is larger than the average $B(s)D_A D(s)F(s,x)$ of the original population, where both F's are evaluated at the density of the original population.

In this model, the loci influencing the size of the organism have pleiotropic effects on density-independent and density-dependent survival and on fecundity due to the influence of size on these ecological parameters. Thus, the phenotype size has partitioned this seemingly complicated genetic situation into a manageable genetic model and a manageable ecological model. So we are free to discuss the ecological constraints on the relation between size and fitness, which appears as an observable relation, rather than the pleiotropy of genes influencing the dynamical parameters of the ecological model, which in principle, but not in practice, appears observable.

As we will study the evolution of offspring size by considering conditions for the initial increase of a new allele, the model may be simplified further. The condition for initial increase of a rare allele clearly shows insensitivity whether the mortality in the population is density-dependent or not. The density-dependent death-rate factor, F, in the conditions is evaluated at the equilibrium density in the original population. Therefore, the factor F is a constant during the process that we study, so it may be absorbed into the survival probability D. This argument pertains to a population in a constant environment, but it extends to a fluctuating environment (Christiansen 1984a). The evaluation of the initial increase condition in a fluctuating environment is done by averaging $\log[B(s)D_A D(s)F(s,x)]$ over an appropriate spectrum of environments and densities. The average for the new genotype and for the original population is made using the same environmental spectrum, as they occur in the same physical and biotic environment. Therefore, without much loss of generality, we may initially consider the more simple situation where the density-dependent death-rate factor, f, is independent of the phenotype.

Size, Survival, and Reproduction

Size has an immediate relation to survival for a multitude of reasons. Within reasonable limits of comparison, small organisms are in general less buffered toward the physical environment than large organisms, and they would seem more likely to fall victim to a predator. So we may expect d and f to be decreasing functions of size, at least for sufficiently simple organisms.

The relation between size and reproduction is considerably more complicated. A large organism can devote more energy to reproduction on an absolute scale, but fecundity is the number of offspring, not the amount of energy. However, given the fecundity and the size of the offspring in units of energy, the reproductive output of an individual can be compared to its size in energy units, and this provide the constraint that size imposes on fecundity. This comparison, however, may be done in a number of ways which provide quite different relations between size and fitness.

Let us consider a simple model where a certain part, E_g, of the energy, E, available to the adult organism is devoted to reproduction. The fraction of reproductive energy may be called the reproductive effort, $e = E_g/E$. The fecundity of the individual now is found by partitioning the reproductive energy into offspring of size s, so:

$$B(E,e,s) = \frac{eE}{s}. \tag{6}$$

This simple model connects the reproductive size, E, and the reproductive effort, e, to the fecundity of the individual in terms of the offspring size, s.

Let us first consider the situation of genetic variation in s and assume each genotype produces offspring of a given size. This introduces dynamical pleiotropy on the fecundity and the survival during development, in that a female parent of a given genotype produces eggs of a size that will determine her fecundity from Eq. (6) and the survival of the offspring from Eqs. (4) and (5). As mating is random, a male parent will fertilize eggs equal to the average size of offspring in the population.

We will analyze the condition for initial increase of a rare allele, so suppose that the population is initially monomorphic aa at an autosomal locus influencing offspring size. Females of genotype aa produce offspring of size s_0. Introduce into this population a new allele A, so females of the new rare genotype Aa produce offspring of a different size, s_1. As the new allele is assumed rare we may neglect the occurrence of genotype AA. By the same vein of approximation the males of genotype Aa will on the average fertilize eggs of a size s_0. In addition, we assume that the size of the reproducing individual and the reproductive effort are the same for the two genotypes, i.e., $E_1 = E_0$ and $e_1 = e_0$, and that the size of the reproducing individual and the reproductive effort are the same for the two phenotypes, i.e., individuals grown from eggs of size s_1 do not differ from individuals grown from eggs of size s_0 at the time of reproduction. With these assumptions the new allele will increase in frequency when:

$$B(s_0)D(s_0) \langle B(s_1)D(s_1) \tag{7}$$

and decrease in frequency when

$$B(s_0)D(s_0) > B(s_1)D(s_1). \tag{8}$$

If either the size of the reproducing individual or the reproductive effort are different for the two phenotypes, then the new allele will introduce maternal effects on fecundity, and the conditions for initial increase become more complicated.

With the above assumptions, i.e., that the genotypic variation only affects the egg size, and that no maternal effects on fecundity occur, the relative fecundity of females of the rare genotype is $B(s_1)/B(s_0) = s_0/s_1$. This relative fecundity is always less than one for $s_1 > s_0$, so fecundity selection favours decreased size of the offspring.

The size-dependent survival (Eq. 4) may be specified either in terms of a constant development time, T, or in terms of a development time which depends on the initial size, s. First consider the situation where the development is terminated when the individual has reached a given size, S (Christiansen and Fenchel 1979). Then, the development time, T, is a function of the egg size, s, in that the development time is determined from the equation $\sigma(T) = S$, so from Eq. (1):

$$T(s) = \int_s^S \frac{1}{\sigma g(\sigma)} \, d\sigma. \tag{9}$$

During development the difference in survival between the two phenotypes depends only on the difference in size during the initial development and not on the length of the period of development (Eq. 9). From Eq. (4) using Eq. (1) we get:

$$\frac{D(s_1)}{D(s_0)} = \exp\left[\int_{s_0}^{s_1} \frac{d(\sigma)}{\sigma g(\sigma)} \, d\sigma \right]. \tag{10}$$

This ratio, the relative viability of the new rare genotype, is larger than unity if and only if $s_1 > s_0$, so zygotic selection favors increased size of the offspring. Thus, the two selection components, zygotic selection and fecundity selection, work in opposite directions, and as the new allele increases in frequency if and only if $B(s_1)D(s_1)/B(s_0)D(s_0) > 1$, the separate evaluations do not help. If the difference in egg size, $s_1 - s_0$, is numerically small, then we can approximate the condition for initial increase to provide an easier evaluation: allele A increases if and only if:

$$\frac{s_1 - s_0}{s_0} \left[1 - \frac{d(s_0)}{g(s_0)} \right] < 0 \tag{11}$$

(Christiansen and Fenchel 1979).

The condition for increase of an allele that changes egg size is, therefore, determined primarily by the relation between the initial death rate, $d(s)$, and the initial specific growth rate, $g(s)$, during development. Thus, if the initial death rate is higher than the initial specific growth rate, then larger egg size will be favored, and conversely, if the initial growth picks up more energy than is lost by death, then smaller egg size is favored. Sigmoid growth means that the specific growth rate will decrease with size and attain its maximum, less than $\alpha = g(0)$, at a very small size. The organism must have a minimum size of the egg that allows development, and this can be introduced into the model by letting $d(\sigma)$ increase to infinity as σ becomes very small. Thus, for σ sufficiently small evolution will always favor increased egg size. For reasonable assumptions on the functional form of d and g used in Fig. 1, this results in evolution for either a very small egg size, s^*, or a very large egg size, S, possibly with the direction of evolution dependent on the initial egg size. In Fig. 1 the simple specific growth rate of Eq. (2) is used, and it is assumed that a given increment in size will have a larger effect on the death rate, the smaller the individual.

To further illustrate the interpretation of condition (7) in terms of the evolution of the egg size, let us consider the situation where the development time, T, is independent of the initial size, s. Zygotic selection again favors increased size. The relative viability of the new rare genotype, $D(s_1)/D(s_0)$, is larger than one if and only if $s_0 \langle s_1$, since the smaller offspring have the highest death rate at any time during development. If the difference in egg size, $s_1 - s_0$, is numerically small, then we can approximately determine that the new allele increases if and only if:

$$\frac{s_1 - s_0}{s_0}\left[1 - \frac{d(s_0) - d(S_0)}{g(s_0)}\right] \langle 0. \tag{12}$$

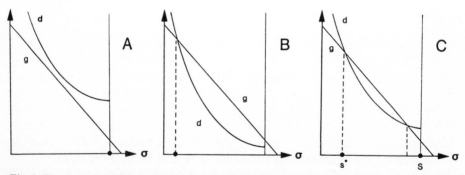

Fig. 1. The three possibilities of evolution of offspring size in terms of the death rate d and the specific growth rate g as functions of the offspring size σ with a fixed size S ending development (condition 11, assuming g given by Eq. 2). **A** Selection for increased offspring size. This is the situation which favors direct development in marine invertebrates. **B** Selection for a small offspring size. This is the situation which favors planktotrophic development in marine invertebrates. **C** Selection for a small or a large offspring size dependent on the initial size of offspring. For marine invertebrates this situation may either lead to planktotrophic development or to direct development (After Christiansen and Fenchel 1979)

This condition is almost identical to condition (11) if the death rate by the end of the development, $d(S_0)$, is small compared to the initial death rate, $d(s_0)$. However, the qualitative properties of condition (11) as they appear in Fig. 1, are sensitive to this change in the condition. In terms of Fig. 1 condition (12) is illustrated by assuming that $d(1/\beta) = 0$. Thus, condition (12) corresponds to the case in Fig. 1A, when the dependence of the death rate on individual size is very strong, i.e., when the curve $d(\sigma)$ is very steep. In this instance, evolution will progress toward a very large offspring size. For more moderate dependence of the death rate on individual size, condition (12) corresponds to the case in Fig. 1B, where evolution progresses toward a small offspring size, s^*, the evolutionary stable offspring size.

The difference between the conditions is smaller if we introduce into the models the differences between the phenotypes in their entrance into the adult stage. Condition (12) should be modified to account for the different initial size of the adult individuals, and condition (11) should account for the longer adult life of individuals from larger eggs. From Eq. (9) we get that the difference in the developmental period and, therefore, supposedly in the length of adult life is

$$T(s_1) - T(s_0) = -\frac{s_1 - s_0}{s_0} \frac{1}{g(s_0)} \tag{13}$$

for $s_1 - s_0$ numerically small. So adjusting condition (11) for the adult death rate, d_A, during this period amounts to substituting d_A for the death rate $d(S_0)$ in condition (12). The situation depicted in Fig. 1C therefore is possible only if the time difference of Eq. (13) is immaterial, i.e., if d_A is clearly smaller than $d(S_0)$. Thus, condition (11) describes the evolution of offspring size when the environment of the developing individual is clearly distinct from the environment of the adult individual.

The size of the egg has the potential of influencing the size of the adult organism. In the model with a size-dependent termination of the developmental period, the initial adult size of the two phenotypes, s_0 and s_1, are equal. Accordingly, it is reasonable to assume that the energy available at the time of reproduction is equal. The two phenotypes, however, spend different times developing (Eq. 9), so it is reasonable to assume that they spend different lengths of times maturing as adults (Eq. 13). This may well give rise to effects of maternal genotype on offspring size at the time of reproduction, and thereby to effects on offspring fecundity. In the model with a constant development period, the offspring of the different phenotypes enter into the adult stage with different sizes, so the possibility for maternal effects seems more immediate. To avoid the complication of maternal effects, we assumed independence between the size of the reproducing individual, E, and the size after development, S, but in general a range of relations between E and S is possible.

To produce an example of an extreme maternal effect assume that the size of the reproducing individual, E, is proportional to the size after development, S. Thus, two sizes, S_0 and S_1, of reproducing individuals exist according to whether they originated from eggs of size s_0 or s_1. A reproducing individual of the rare genotype Aa will have the size S_0 if its father had the genotype Aa, and the size S_1 if its mother had the genotype Aa (the approximation assuming that the genotype is rare neglects the possibility that both parents carry the allele A). Then we can assume that the fecundity of the two phenotypes among the Aa females, S_1 and S_0, relative to the fecundity of the aa females is $B(S_1,s_1)/B(S_0,s_0) = S_1 s_0/S_0 s_1$ and $B(S_0,s_1)/B(S_0,s_0) = s_0/s_1$, respectively. With sigmoid growth these relative fecundities are both greater than one when $s_0 \rangle s_1$, and again fecundity selection will favor decreased size. The exact condition for the increase of the new allele is a bit more complicated than condition (7), but if the difference in egg size, $s_1 - s_0$, is numerically small, then we get approximately that the new allele increases if and only if:

$$\frac{s_1 - s_0}{s_0} \left[1 - \frac{d(s_0) - d(S_0) + g(S_0)/2}{g(s_0)}\right] \langle 0. \tag{14}$$

This condition is almost identical to condition (12) if $g(S_0)$ is small compared to $g(s_0)$, i.e., if the growth curve is very sigmoid. This is not surprising, as very sigmoid growth produces very small maternal effects on fecundity. However, what is of main concern here is, that the qualitative properties of the model are again shown by Fig. 1A and B. The evolution of the offspring size is predicted to have the same pattern as in the model with no maternal effect, but maternal effects will clearly increase the evolutionary stable offspring size, s^*, and relax the condition for evolution to progress toward a very large offspring size.

Christiansen and Fenchel (1979) developed the model of the evolutionary interplay of fecundity and survival of offspring as an extension of a model of Vance (1973). The arguments were used to explain the occurrence of two distinct reproductive strategies in marine invertebrates, one (planktotrophic development) with very small offspring size and the other (direct or lecithotrophic development) with much larger offspring size corresponding to the metamorphosis size of planktotrophic larvae. The metamorphosis constitutes a clear discontinuity between the period of development of the planktotrophic larvae and later stages in the life cycle of the organism (so the robustness assumption, that d_A and $d(S_0)$ are very different, seems appropriate in this instance). This dichotomy in life history exists as variation among species at lower taxonomic levels within many higher taxa, so the change from one to the other way of reproduction must have occurred repeatedly. The evolutionary instability predicted by the model would produce exactly this kind of variation among species. In addtion, the model can explain more subtle variations. The latitudinal cline in the frequency of planktotrophic development in, for instance, marine bivalves (Ockelmann 1965), where planktotrophic development becomes rarer with increasing latitude, is predicted from the model under the crude assumption that growth rate decreases (or mortality increases) with latitude. The slower growth in colder waters produces a lower probability of survival for a given egg size even with unchanged death rates for given sizes. If this explanation is correct, then the model further predicts that species maintaining planktotrophic development in the north will produce larger eggs.

If this explanation of the decrease of egg size with temperature is correct among species with planktotrophic development, then we will expect a similar pattern among species where the development to adulthood is smoother. In accordance with this a decrease of egg size with temperature was also observed in species with direct development by Clarke (1979) in a latitudinal comparison of decapods and by Kolding and Fenchel (1981) in a comparison between summer- and winter-breeding gammarids.

The latitudinal cline in the mode of reproduction may also be viewed as a consequence of a more unpredictable environment in the north. The condition (7) for the initial increase of a rare allele is in a varying environment replaced by the similar condition on the geometric average of $B(s)D(s)$ through time (Haldane and Jayakar 1963; Gillespie 1973). In condition (11) this corresponds to replacing the ratio of the death rate and the specific growth rate, $d(s_0)/g(s_0)$, by the (arithmetric) average value of this ratio over time. Therefore, in a varying environment condition (11) is interpreted by considering the average of the ratio $d(\sigma)/g(\sigma)$ over time, so allele A increases if the condition is on the average fulfilled. Now, if the environmental variation is in the death rate, leaving the growth rate constant, then condition (11) is a comparison of the average death rate and the specific growth rate, so the latitudinal cline may emerge for the same reason as in the constant environment model. If the environmental variation is in the growth rate only, then (by Jensen's inequality of expected values, see e.g., Karlin and Taylor 1975, p. 249);

$$E\left(\frac{d(\sigma)}{g(\sigma)}\right) \rangle \frac{d(\sigma)}{E(g(\sigma))} , \tag{15}$$

where $E(\cdot)$ designates the average over time. Thus, variation in the growth rate will mimic the effect of a lower growth rate, when compared to a constant environment with the average growth rate. In the simple growth model (2) with moderate variation in the growth rate parameter α, this effect will approximately produce a simple effect of the variance of the parameter, in that

$$E\left(\frac{d(\sigma)}{g(\sigma)}\right) \approx \frac{d(\sigma)}{E(g(\sigma))} \left[1 + \frac{Var(\alpha)}{E(\alpha)^2}\right], \tag{16}$$

where $Var(\cdot)$ designates the variance over time. If the death rate and the specific growth rate show positive covariation, then the effect of the variation in the specific growth rate is diminished. However, the qualitative importance of environmental variation is small, when the latitudinal cline is of concern. Variation in the size-dependent death rate tends to be averaged out, and increased variation in the growth rate with latitude has the same effect as a decrease in the mean growth rate. Thus, we need only realize that the two considered changes in the growth of offspring with latitude are reflections of the same link between the growth of the individual and the environment in which it grows.

The dichotomy of reproductive behaviour in marine invertebrates has consequences for many other aspects of evolution (Strathmann 1985). Planktonic larvae necessarily are dispersed more than offspring with direct development, so the genetic structure of populations will differ among species dependent on the way of reproduction. Further, the presumed differences in environmental susceptibility of the offspring of different sizes will also produce differences in the population dynamics of the species, in that the recruitment in species with planktotrophic development is expected to vary more with the environment than recruitment in species with the larger, direct developing offspring. In a varying environment organisms with small offspring should therefore show variation in the success of reproduction as measured by the recruitment after development. Thus, temporal instability in population size is in this way more a consequence than a cause for the evolution of offspring size (Christiansen and Fenchel 1979).

Reproductive Effort and Survival

The fecundity of an adult individual can be changed with no immediate effect on the individual by changing the offspring size while keeping the reproductive effort constant (eq. 6). However, fecundity may also change as a function of the reproductive effort of the individual, and as this involves changes in the energy allocation of the individual, it must involve changes in its probability of survival. For instance, in a study of the life histories of four ascidian species Svane (1983) concluded that increased investment in the protective tunic leads to increased longevity and to lower fecundity.

Genotypic variation in reproductive effort introduces dynamical pleiotropy on the fecundity and the survival during the adult stage, in that a parent of a given genotype allocates reproductive energy which determines her fecundity from Eq. (6), but this allocation must also influence her probability of survival during the adult stage. Therefore, we may assume that the adult survival, $D_A(e)$, is a decreasing function of the reproductive effort, e;

Let us therefore as before seek the condition for initial increase of a new allele, C, say, at a locus, that influences the character reproductive effort, and that starts out monomorphic for allele c. Each genotype is given a reproductive effort, e_0, say, for genotype cc in the original population and e_1 for the new rare genotype Cc. We assume each offspring to have a given size, s, independent of the maternal genotype at the considered locus. The new allele will then increase in frequency when

$$B(e_0)D_A(e_0) < B(e_1)D_A(e_1). \tag{17}$$

Again, the fecundity and viability factors react differently to changes in the phenotype. The fecundity B is an increasing function of e, and the survival D_A is a decreasing function of e. However, there must be a limit to the amount of assimilated energy that can be allocated to reproduction. This constraint can be modeled by letting the adult death rate, d_A, increase to infinity (that is, D_A decreases to zero) as e approaches unity. If the change in reproductive effort is small, i.e., $e_1 - e_0$ is numerically small, then condition (17) becomes approximately

$$\frac{e_1 - e_0}{e_0} [1 - (1-T)e_0 \frac{dd_A(e_0)}{de}] > 0 \tag{18}$$

(the generation time is set to one time unit, so $1 - T$ time units elapse after development).

It is natural to assume that $d_A(e)$ increases from the reproduction-independent death-rate level, $d_A(0)$, to infinity as e approaches unity. However, this assumption guarantees only that the derivative of d_A in condition (18) is positive. Instead, we may assume that the higher the initial reproductive effort is, the more serious is the effect that a given increment in reproductive effort has on the survival of the individual. This amounts to assuming that the derivative of the adult death rate, dd_A/de in condition (18), is an increasing function of the reproductive effort, e. With this assumption, the equation:

$$(1-T)e_0 \; \frac{dd_A(e_0)}{de} \; = 1 \tag{19}$$

has a unique solution, e^*. This solution is a critical reproductive effort, in that the sign of the second factor in condition (18) depends on whether the reproductive effort is above or below the solution to Eq. (19).

The relation between e_0 and e^* now provides a simple characterization of the evolution of the reproductive effort. Alleles that increase the reproductive effort, $e_1 \rangle e_0$, will increase in frequency when the initial reproductive effort is lower than the critical reproductive effort, i.e., $e_0 \langle e^*$, and decrease when the initial reproductive effort is higher than the critical reproductive effort, i.e., $e_0 \rangle e^*$. Thus, the critical reproductive effort, e^*, is an evolutionary stable strategy, in the sense that the population tends to evolve toward that reproductive effort, and that an allele which influences reproductive effort will not increase if introduced as rare in a monomorphic population with the reproductive effort equal to e^*.

Data on reproductive effort in marine invertebrates are scattered and incomplete. However, the above cited investigations of Clarke (1979) and Kolding and Fenchel (1981) include observations to illustrate variation among species in reproductive effort. The latitudinal comparison of Clarke (1979) showed that the reproductive effort of arctic species is higher than that of temperate species. In terms of the analytical result this is consistent with a steeper increase in the adult death rate, d_A, with reproductive effort, e, i.e., the increase in d_A starts at lower values of e for arctic species. In the gammarids (Kolding and Fenchel 1981) a similar simple pattern in variation with temperature is not apparent. However, these data show the parallel and consistent geographical variation among species that is expected in a character subject to stabilizing evolutionary forces (Fig. 2). Four of the investigated species show very similar reproductive efforts in an estuary in Denmark, and they show the same similarity in the Baltic, but the reproductive effort is uniformly higher in the Baltic than in the Danish estuary. The one species that has a different reproductive effort shows a shift comparable to the other species when Baltic and Danish populations are compared. An investigation of two of the species of the uniform group along the Atlantic coast of Western Europe shows a higher level of reprodutive effort than at other places, but again the species show very similar reproductive effort.

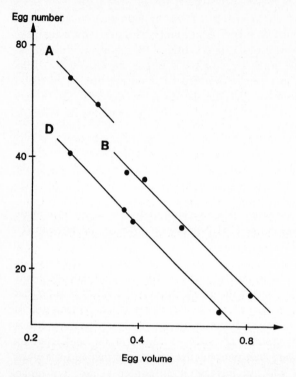

Fig. 2. The variation in reproductive effort in four species of the genus *Gammarus* in three geographical areas. The *dots* depict for a species in a given place the relationship on a double long scale between the volume of an egg (in mm³) and the average number of eggs in broods from females of 10 mm size. The *lines*, drawn with slope −1, collect the *dots* corresponding to the species in a given area. *A* Atlantic coast of France; *B* Baltic Sea; *D* a Danish estuary. The two species observed in area *A* are *G. locusta* and *G. salinus.* The two additional species observed in areas *B* and *D* are *G. zaddachi* and *G. duebeni.* The lines connect points with the same product of egg volume and egg number, so they connect points with the same egg biomass. The egg biomass of a standard size female seems as good a measure of reproductive effort as any. Thus, we may conclude that the reproductive effort is very similar for the species in the same geographical area. In addition, parallel geographical variation in reproductive effort exists, in that area *D* has the lowest and area *A* the highest reproductive effort (after Kolding and Fenchel 1981). This investigation also included the species *G. oceanicus* in areas *B* and *D*. This species is different from the other species, but what is of concern here is that its reproductive effort is larger than the reproductive effort of the other four species in both areas. However, the difference in reproductive effort between the two areas is comparable to the difference shown in this figure (Kolding and Fenchel 1981)

Discussion

The dynamical pleiotropy induced by variation in egg size may produce natural selection which results in diversifying evolution, in that related species occupying relatively similar environments may have very different phenotypes as a result of minor differences in the environment. On the other hand, the dynamical pleiotropy induced by variation in reproductive effort results in a stabilizing evolution, in that minor differences in the environment of related species will result in minor differences in reproductive effort. These two very simple characters, that connect the fecundity of an individual to the survival of individuals in the life stages immediately important to reproduction, therefore, may produce very different evolutionary results. The two different results may be viewed as extremes in a model where the dynamical pleiotropy of the genotypes are modeled directly, but within this framework the result would be the trivial statement that anything can happen.

The extreme contrast between the two characters is only seen in situations where a clear distinction between the developmental stage and the adult stage exists. However, if this is not the case, the difference in evolutionary pattern between the two characters is still significant. Evolution of offspring size depends directly on the magnitude of the size-dependent death rate, in that the direction of evolution is determined by a comparison to the size-dependent specific growth rate. Evolution of reproductive effort depends only on the way the adult death rate changes with reproductive effort, not on the magnitude of the death rate. Thus, a change in the environment that changes the death rate at all life stages will have an immediate effect on the natural selection working on offspring size, but it will only have an effect on the natural selection working on reproductive effort, if it changes the way in which the adult death rate varies with reproductive effort.

The differences in the pattern of evolution of the two kinds of reproductive characters considered here are expected to be reflected in the pattern of data comparing the reproduction of different species. Offspring size is expected to show a larger variance than reproductive effort, so covariance studies of offspring size in relation to the environment or to the behavior of the organisms are more likely to show a pattern. Furthermore, offspring size is easier to quantify than reproductive effort. The dichotomous pattern in offspring size, therefore, is expected to dominate comparative studies of reproductive characters.

We assumed that the density-dependent death-rate coefficients were the same for all genotypes. For arguments based on the increase of initially rare alleles this assumption is immaterial, as the death rate is evaluated at the prevailing density of the population which is not changed by the presence of a rare allele. If a new allele increases, it will eventually influence the dynamics of the population and the variation in density-dependent death rates becomes relevant. However, as long as the density-dependent death rate of all genotypes in the population is determined by the total density of the population, the distinction between density-dependent and density-independent death rate is immaterial for evolutionary arguments in a logistic model. Density dependence becomes important in evolution when the sensitivity of a given genotype to density varies as a function of the genotypic composition of the population, as, for instance, when the phenotypes in question show differences in their utilization of a limiting resource. Such differences may relax the condition for initial increase and result in the establishment of a stable polymorphism. Therefore, they constitute complications which are of more immediate interest in population genetics than in population ecology.

The simple models considered here may be compared to trade-off models of evolutionary ecology giving rise to models involving trade-off pleiotropy of the considered genetic variation (in the sense of Prout 1980). These models are in the tradition of population genetics, in that the genotypic effects on the primary fitness components are modeled, so the trade-off effects are modeled as negative correlations between the fitness components. All the situations that we considered here result in such models. However, as stressed before the pleiotropic effects on fitness of a genotype is inferred from its effect on one particular phenotype of the individuals. Thus, in the classical sense of the concept we actually neglect the complication of pleiotropy, and assume that the considered genetic variation only affects the particular phenotype of interest. The most lucid expression of this is the way in which the fitness component of fecundity is partitioned into a variety of attributes of the individual to ease the evaluation of the trade-off effects of the phenotypic variation. This is appropriate in many considerations of simple organisms, as the effect of reproduction on the individual is mediated by anything else than the cold number expressing fecundity.

The results on the evolution of offspring size and reproductive effort clearly contradict the arguments that large offspring size or small reproductive effort develops as a response to a competitive environment and that small offspring size or large reproductive effort develops as a response to a high and unpredictable density-independent mortality during initial development or to a low and unpredictable rate of reproductive success (for a review see Stearns 1976). These well-known arguments on r- and K-selection are based on results obtained in what I called evolutionary ecology models in the tradition of population genetics, in which genetic variation in the parameters of the ecological model is studied. These results are generalizations of the principle of fitness maximalization to the situation of density-dependent fitnesses, where it entails maximization of the equilibrium population size (Fisher 1930, 1958; MacArthur 1962; Anderson 1971; Roughgarden 1976). The interpretations of this in relation to the r-K scenario is founded on the assertion that the equilibrium population size in an undisturbed population is an important biological parameter (Boyce 1984). If alternatively density-dependent natural selection if formulated in terms of density-dependent birth and death rates, then the principle of fitness maximization is virtually the same for density-dependent and density-independent natural selection (Fisher 1930, 1958; Kostitzin 1936, 1938), with the result that the qualitative distinction between r- and K-selection as theoretical concepts vanishes (Christiansen 1984c).

The results of the population-genetics approach are not necessarily less biologically relevant than the result reached by the present modeling approach. However, the nature of the dynamical pleiotropy of the considered genetic variation is not expliticitly modeled, and the present results show that the arguments relating offspring size to a competitive or not competitive early environment contains hidden assumptions which may be relevant when understanding the variation among species.

Summary

Phenotypic variation that produces variation in the relation between individuals and their environment is both an ecologically and evolutionary important variation. Within the framework of evolutionary ecology natural selection and variation in individual fitness is often described as individual variation in dynamical parameters of ecological models. The results of the study of such models are general statements concerning the evolutionary push on important dynamical parameters. Alternatively, the variation in a morphological character may be considered as the primary phenotypic variation. Then a model for the study of the evolution of this character requires a simple description of the genetics of the character, on the one hand, and a simple ecological description of the character, on the other hand. Such a modeling approach produces interesting evolutionary models of the interplay between different parameters of dynamical models in ecology. Examples of the use of this approach include the interplay between body size, reproductive effort, and fecundity.

Acknowledgments. The development of the analysis of evolution of reproductive effort was done in collaboration with Dr. Tom Fenchel, and the presentation benefited from discussions with him. The figures were drawn by Mr. Arno Jensen. The research was supported in part by grant 81-5458 from the Danish Natural Science Foundation.

References

Anderson WW (1971) Genetic equilibrium and population growth under density-regulated selection. Am Nat 105:489–498

Boyce MS (1984) Restitution of r- and K-selection as a model of density-dependent natural selection. Annu Rev Ecol Syst 15:427–447

Christiansen FB (1984a) Evolution in a temporally varying environment: density and composition dependent genotypic fitnesses. In: Wöhrmann K, Loeschke V (eds) Population biology and evolution. Springer, Berlin Heidelberg New York, pp 115–124

Christiansen FB (1984b) The definition and measurement of fitness. In: Shorrocks B (ed) Evolutionary ecology, BES Symp 23. Blackwell Sci, Oxford, pp 65–71

Christiansen FB (1984c) Natural selection related to the biotic environment. In: Jayakar SD, Zonta L (eds) Evolution and the genetics of populations. Atti Ass Genet Ital XXIX (Suppl), pp 85–102

Christiansen FB, Fenchel TM (1979) Evolution of marine invertebrate reproductive patterns. Theor Popul Biol 16:267–282

Clarke A (1979) On living in cold water: K-strategies in Arctic bentos. Mar Biol 55:111–119

Darwin C (1872) The origin of species, 6th edn. John Murray, London

Feldman MW, Christiansen FB, Liberman U (1983) On some models of fertility selection. Genetics 105:1003–1010

Fisher RA (1930) The genetical theory of natural selection. Claredon, Oxford

Fisher RA (1958) The genetical theory of natural selection, 2nd edn. Dover, New York

Gillespie J (1973) Polymorphism in random environments. Theor Popul Biol 4:193–195

Haldane JBS, Jayakar SD (1963) Polymorphism due to selection of varying direction. J Genet 58:237–242

Karlin S, Taylor HM (1975) A first course in stochastic processes. Academic Press, London

Kolding S, Fenchel TM (1981) Patterns of reproduction in different populations of five species of the amphipod genus Gammarus. Oikos 37:167–172

Kostitzin VA (1936) Sur les équations différentielles du problème de la sélection mendélienne. C R Acad Sci Paris 203:156–157

Kostitzin VA (1938) Equations différentielles générales du problème de sélection naturelle. C R Acad Sci Paris 206:570–572

MacArthur RH (1962) Some generalized theorems of natural selection. Proc Natl Acad Sci USA 48:1893–1897

Ockelmann KW (1965) Development types in marine bivalves and their distribution along the Atlantic coast of Europe. In: Cox LR, Peake JS (eds) Proceedings of the first european malacological congress. Malacological Society, London, pp 25–35

Poulsen ET (1979) A model for population regulation with density- and frequency-dependent selection. J Math Biol 8:325–343

Prout T (1980) Some relationships between density-independent selection and density-dependent selection. Evol Biol 13:1–68

Roughgarden J (1976) Resource partitioning among competing species – a coevolutionary approach. Theor Popul Biol 9:388–424

Stears S (1976) Life-history tactics: a review of the ideas. Q Rev Biol 51:3–47

Strathmann RR (1985) Feeding and nonfeeding larval development and life-history evolution in marine invertebrates. Annu Rev Ecol Syst 16:339–361

Svane I (1983) Ascidian reproductive patterns related to long-term population dynamics. Sarsia 68:249–255

Vance RR (1973) On reproductive strategies in marine benthic invertebrates. Am Nat 107:339–352

Chapter 7 **Constraints in Selection Response**

W. Scharloo [1]

In evolutionary biology two extreme, opposite opinions are presently found:
1. Some ecologists cultivate the evolution of life histories implying that all that matters in evolutionary ecology is optimization. This approach takes for granted that for all characters of ecological importance there is plenty of genetic variation, expressed as continuous phenotypic variation, permitting gradual evolutionary change under the influence of natural selection.
2. Some paleontologists claim that evolution is not the smooth approach to perfect adaptation which they see to be the trademark of neo-Darwinism. They launched the theory of punctuated equilibria: evolution would proceed in short bursts of change associated with species formation. Most of the time species would be in a state of internal equilibrium which would cause the long periods of stasis which species often show in the paleontological record. The direction of change would be caused to a large extent by developmental constraints.

This, in fact, contradicts the idea that the morphological change associated with species formation is essentially random and that species selection causes the long-term direction of evolution.

Supporters of the role of developmental constraints in evolution, in particular during the long periods of stasis in evolution, have stated that quantitative genetics has generated evidence that supports this theory (Alberch 1980, 1982).

The presence of developmental constraints is often suggested by the study of variation patterns of quantitative characters in genetically different populations of the same species and of related species.

However, the only direct experimental approach to analyse the role of developmental constraints is trying to change quantitative characters by artificial selection and/or environmental manipulation.

Developmental constraints can be defined as bias in production of variant phenotypes or limitations on phenotypic variability caused by the structure, character, composition or dynamics of the developmental system.

---

[1] Department of Population and Evolutionary Biology, University of Utrecht, The Netherlands

Genetic Constraints on Adaptive Evolution
Ed. by V. Loeschcke
© Springer-Verlag Berlin Heidelberg 1987

Quantitative Genetics

Let us see what quantitative genetics has shown us in this context. In *Drosophila* a large number of experiments with artificial selection have been performed on a large variety of characters and were almost always successful.

The pattern of response in the classical experiments performed with *Drosophila* on artificial selection on bristle number and body size (see Falconer 1981) is always an immediate reaction lasting 20 generations in which a plateau is reached. Then no further change occurs notwithstanding continued further selection. It has been argued that these plateaus demonstrate the importance of developmental constraints (Alberch 1980). However, analysis of plateaued populations have invariably shown that this is not the correct explanation: plateaus are a consequence of exhaustion of genetic variation suitable to sustain further response (e.g. Clayton and Robertson 1957).

This is very clearly shown by an experiment by Yoo (1980) on sternopleural bristle number in stocks homozygous for the *scute* mutant. In the base population the mean value is near eight *in scute*. In other experiments the usual plateaus were found, but then the selection populations were rather small, let us say not more than 20 individuals. Yoo used five times as many individuals and obtained a continuous smooth response with no indications of plateaus even after 90 generations whereby he obtained a mean bristle number of more than 40.

Gene-Environmental Factor/Phenotype Mapping Functions (GEPM)

Characters such as body size, sternopleural bristle number and abdominal bristle number showed no signs of irregularities in the expression of variation. The scale of the characters is linear, at least on a log scale. Then we can make a graphical model of the expression of genetic variation and the effects of environmental change, i.e., plot a gene-environmental factor/phenotype mapping function (GEPM) (Fig. 1).

We assume that there are normal distributions of genetic variation in the first instance expressed as normal distributions of gene products or morphogenetic substance. They are in the case of a linear mapping function expressed as normal distributions of phenotypes. When such a distribution of gene products is changed by artificial selection on phenotypes, the phenotypic distributions shift in a regular, continuous fashion.

In the late 1950's several investigators were looking for situations in which the mapping functions were not linear, in which the expression of genetic variation was biased, in which the developmental processes connecting genotypes and phenotypes were constrained or − to use Waddington's terminology − canalized. These investigators were inspired by Waddington's book *The strategy of the genes* (1957). An important topic in this book was his emphasis on the relatively small variability of wild-type phenotypic characters when compared with the variability of the same characters in mutants and in abnormal phenotypes induced by abnormal environ-

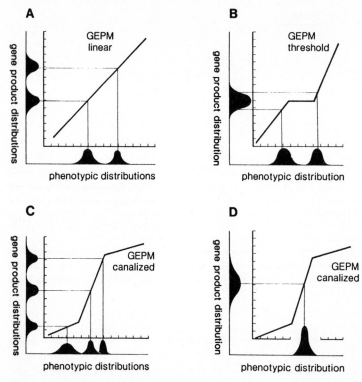

Fig. 1. Model for the effect of the shape of gene-environmental factor/ phenotype mapping functions (GEMPs). On the ordinates normal distributions of gene products (= morphogenetic substance, in other figures in this chapter bristle-forming or vein-forming substance). The variation is caused by genetic variation, variation of environmental factors and stochastic variation (developmental error, developmental noise; resulting in differences between characters in the same individual, e.g. the two wings of an individual fly). The mean of the distribution can be shifted by a major mutant, by artificial selection or by environmental factors. The normal distributions of the morphogenetic substance are translated by the mapping function in phenotypic distributions. The shape of the mapping function is determined by developmental processes between the gene product and phenotypic expression, e.g. in bristle number or vein length. When the GEPM is linear (**A**) shifting the normal distribution will be reflected in a linear shift of the phenotypic distributions. When the GEPM is a threshold function, unimodal frequency distributions of gene products can result in bimodal phenotypic frequency distributions (**B**). When the GEPM is sigmoid, the distance between frequency distributions of the gene products is translated in relatively short distances in the steep part of the curve rather than beyond its range (**C**) and variability is far smaller when the gene product distribution is translated in the steep part of the GEPM rather than in the other parts (**D**)

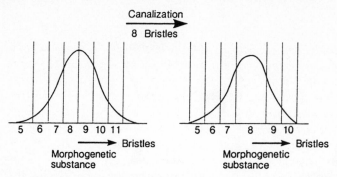

Fig. 2. Variation of a meristic character (i.e. a character that can only be represented in discrete numbers, e.g. egg production or bristle number) generated by continuous variation of a morphogenetic substance and thresholds between the classes of bristle number. When there is a linear scale, the distances between the thresholds separating the bristle classes are equal. When the bristle class 8 is canalized as a consequence of stabilizing selection, the distance between the thresholds enclosing the bristle 8 class is larger than between the thresholds spanning the adjacent bristle classes. Assuming the presence of a normal distribution of morphogenetic substance, the distance between the thresholds can be derived from the frequency of the bristle numbers in the population (see Rendel 1959)

mental factors. His thesis was that wild-type characters, i.e. characters which had a long evolutionary history *under pressure of natural selection of the stabilizing type,* would be canalized. Canalization means that in the wild-type range of phenotypic expression, effects of both genetic and environmental factors would be buffered by properties of the developmental system producing the character involved and would have small phenotypic effects compared with effects in ranges of expression outside the wild-type range.

These gene-environmental factor/phenotype mapping functions (GEPMs) are most easily presented for continous quantitative characters, but this kind of model can also be used for meristic characters which are only expressed in discrete numbers, e.g. number of vertebrae in fish or bristle number in *Drosophila.*

Then we assume an underlying continuous quantitative variable. Transgression of a threshold value then adds or substracts a vertabra or a bristle (Fig. 2). The distance between two thresholds expressed in standard deviations of the underlying variable is often called the width of the class: it shows how much change is necessary to change the number of bristles in the regions involved.

Dunn and Fraser (1959); Rendel (1959); Maynard Smith and Sohndi (1960) and Scharloo (1962) all used such models to explain non-linear gene-environmental factor/phenotype mapping functions (GEPMs).

Scutellar Bristles

Rendel (1959) worked on the number of scutellar bristles in *Drosophila melanogaster*. In the wild-type the number is four and it is almost invariable. Only rarely individuals are found with five or three bristles. In fact, the number of scutellars is a diagnostic character for the genus *Drosophila*. Does this constancy mean that there is no genetic variability present for this character in natural populations? No, when we apply artificial selection for a higher or a lower bristle number by sustained breeding of flies with five or three bristles respectively, we will succeed in changing the bristle number. At first, change will be slow but after a while the rate of phenotypic change per generation will increase. As the mean diverges from the wild-type number and the rate of change increases, the variability too will increase. After some 10–15 generations the mean will be increased to 15 bristles and the bristle number will vary between 8 and 30.

In addition, introduction of the mutant *scute* in a wild-type stock decreases mean bristle number in ♀♀ to two and in ♂♂ to one. The genetic difference between sexes was expressed in this phenotypic range in a difference in bristle number. Moreover, *scute* flies varied. Artificial selection for higher bristle number in *scute* flies succeeded in increasing bristle number, but when approaching the wild-type number four, progress slowed down and it was impossible to obtain *scute* flies with more than four bristles.

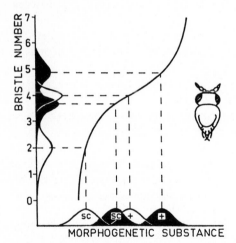

Fig. 3. Model for the formation of scutellar bristles in *Drosophila*. The number of scutellars is supposed to be dependent on the amount of a morphogenetic substance. The sigmoid GEPM causes the canalization of bristle number in the class of four bristles. The morphogenetic substance is supposed to vary between individuals, both in flies carrying the mutant *scute (sc)* and in wild-type flies *(+)* segregating in the same population, according to a normal distribution *(horizontal axis)*. Selection for a higher bristle number in this population shifts both distributions to higher values (distributions in *black*). Before selection the phenotypic variation in *scute* individuals *(vertical axis)* is larger than in wild-type flies; after the selection, when the *scute* flies approach the value of four, they have a small variability. In this selection line the variability of the non-*scute* flies will increase as they pass out of the four-bristle class

Rendel explained these phenomena by designing a model (Fig. 3) in which the mapping function GEPM is not linear, but sigmoid. When gene product (= morphogenetic substance) is plotted on the horizontal axis and phenotypes on the vertical axis, there is a horizontal part in the sigmoid curve, i.e. a zone of no phenotypic change around the wild-type phenotypic value of four notwithstanding increase in morphogenetic substance. Beyond the range of the horizontal part in the sigmoid mapping function phenotypic change is relatively easy.

Vein Length in ci^D

In ci^D *(cubitus interruptus dominant)* expression several criteria pointed to a threshold-like mechanism biasing the expression of both genetic variability and environmental factors (Scharloo 1962). The phenotypic character is here the relative length of the fourth wing vein (Fig. 4). The mutant ci^D causes a terminal interruption of the fourth wing vein. The small fourth chromosome with the ci^D mutant was introduced into the background of three wild populations without disturbing the rest of the genome. In the base populations the phenotypic distributions were unimodal and had more or less normal distributions. The following phenomena point to a non-linear mapping function:

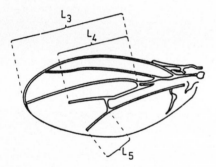

Fig. 4. Measurement of the relative length of the fourth and the fifth vein in experiments with populations carrying the mutants *cubitus interruptus dominant (ci^D), cubitus interruptus dominant of Gloor (ci^{D-G})* or *Hairless (H)*. All three mutants are homozygous lethal and cause interruptions of the fourth and fifth wing vein. The length of the fourth vein was measured as a percentage ratio of the length of the fourth vein distal to the first cross vein (L_4) to the length of the third vein. The fifth vein length was measured as the ratio of the fifth vein distal to the second cross vein (L_5) to the length of the third vein (L_3). Measuring relative length is necessary because there is a considerable change of wing length when the rearing temperature is changed: at lower temperature, wings are longer

1. In all three lines artifical selection for a longer fourth vein generated bimodal frequency distributions in the same range of phenotypic values. After further selection the lower mode was eliminated and the higher mode shifted to still higher values (Fig. 5).

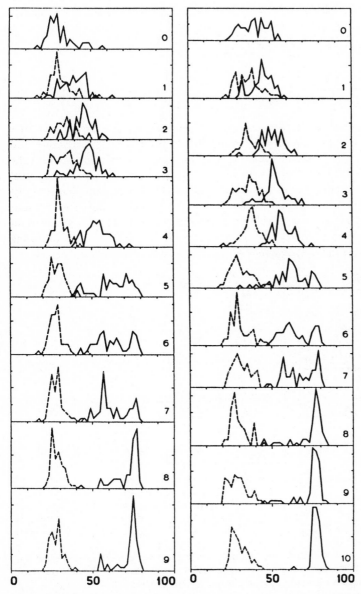

Fig. 5. Frequency distributions in ♀♀ (*left*) and ♂♂ (*right*) of a selection line selected for a larger relative length of the fourth vein in a population with the mutant ci^D. *Horizontal axis:* relative length of the fourth vein; *broken lines:* selection for a short vein; *continuous lines:* selection for a long vein. In both ♀♀ and ♂♂ in the line selected for long veins, bimodal frequency distributions appear when the relative vein length pass through the region 60–80

2. In back-selection lines the bimodal distribution appeared in the same phenotypic region (Fig. 6).

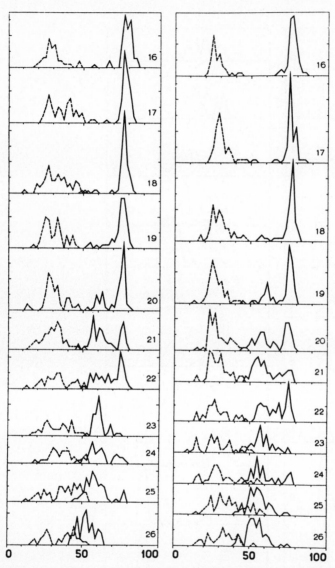

Fig. 6. Frequency distributions of lines selected backwards from generation 15 of the lines depicted in Fig. 5. When the back-selection lines from the high line pass through the region 80–50, bimodal frequency distributions appear again. ♀♀ *left,* ♂♂ *right*

3. When the fourth vein length was changed by temperature (veins become longer at lower temperatures), bimodal distributions appear in the same phenotypic range as in the selection lines (Fig. 7). This occurs both when vein length in line with shorter veins is passing through this range at lower temperatures or when vein length is changed in line with longer veins at higher temperatures.

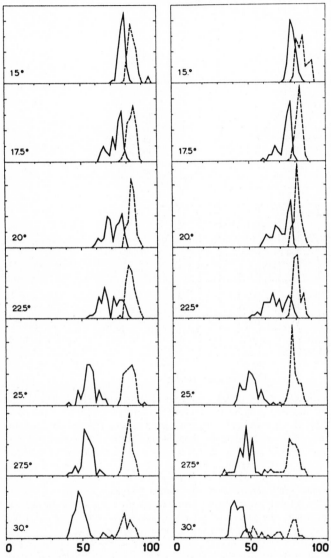

Fig. 7. Frequency distributions in ♀♀ (*left*) and ♂♂ (*right*) of the relative length of the fourth vein at different temperatures. *Continuous lines:* Back-selection lines from the high line, generation 26; *broken lines:* the line selected for a long fourth vein, generation 38. In the back-selection line bimodal frequency distributions occur at 22.5°C and at 20°C when the frequency distributions pass through the vein length region 60–80. In the high lines only part of the ♂♂ fall in this region, which leads to a large variability

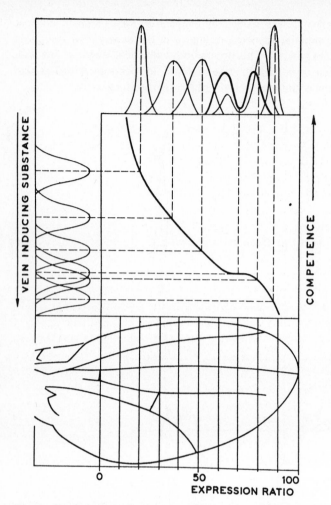

Fig. 8. A model of the formation of the fourth wing vein in *Drosophila*. On the *left* vertical axis normal distributions of the gene product (= vein-inducing substance). They are translated by a non-linear GEPM in the phenotypic frequency distributions at the *top*. The GEPM is generated by a gradient of competence of the cells along the track of the fourth vein to react with vein formation at a certain level of the vein-inducing substance in the wing. A gradient of competence decreasing in a proximal-distal direction explains that the interruption of the fourth wing vein in ci^D is always a terminal interruption. The shape of the GEPM determines the shape of the phenotypic frequency distributions generated from the normal frequency distributions of the morphogenetic substance. The phenotypic frequency distributions and the shape of the gradient of competence are connected with the pattern of wing venation at the *bottom* of the figure. The expression ratio is the relative length of the fourth wing vein as defined in Fig. 4

4. In lines in which vein length is passing through this range of phenotypes by temperature differences there is then a threshold-like relation between temperature and mean vein length.

5. When the lines are passing through this region either by selection or by temperature change there is a large increase in asymmetry, i.e. flies occur frequently which one wing a long fourth vein and in the other wing a short fourth vein.

All these phenomena point to a non-linear gene-environmental factor/phenotype mapping function in which there is a threshold zone in a fixed range of phenotypic values. This was represented in a model (Fig. 8) which shows similarities with Rendel's model for scutellar bristle number. Also in the ci^D model there are hypothetical, underlying normal distributions of a morphogenetic substance or vein-forming substance. But while Rendel's model invokes only the relation between an amount of morphogenetic substance and bristle number, in the ci^D model there is an additional aspect in that the mapping function is related to developmental processes in a specified position in the vein pattern of the wing. As a consequence, the shape of the gene-environmental factor/phenotype mapping function is seen here as the consequence of the shape of a hypothetical gradient of competence of the wing tissue along the track of the fourth vein. Here, the competence means the competence to react to concentrations of the vein-forming substance. This gradient is supposed to decrease from the wing base to the wing tip which could explain that the interruption of the fourth vein is always at the distal end of the vein, i.e. at the wing tip. However, the competence does not decrease linearly from wing base to wing tip, but is supposed to have a region of threshold-like change at the region where bimodal frequency distributions are formed.

Mutants Affect Mapping Functions

The question now arises: are these non-linear mapping functions inescapable, fundamental properties of the structure of the developmental systems of the characters involved? Or are they as Waddington supposed, evolved under pressure of stabilizing natural selection? This question was first posed by Fraser and Kindred (1960), when they discussed a non-linear mapping function found for vibrissae number in mice.

When non-linear mapping functions are evolved under the impact of stabilizing selection and are not the consequence of the fundamental structure of developmental systems, one would expect the presence of genetic variability and the occurrence of mutants affecting the shape of the mapping function.

We saw that in ci^D change becomes more difficult when the length of the fourth vein approaches the wild-type, i.e., when the vein interruption becomes smaller, and that there is a threshold zone when approximately 70% of the vein is present.

However, in the ci^D allele ci^{D-G} (cubitus interruptus dominant of Gloor) the mapping function is linear over a large part of its possible range, e.g. from 30% of the vein present to wild-type.

Moreover, in *Hairless* we found the opposite of the situation in ci^D (Scharloo 1964a and unpublished); there, change is easier when the wild-type is approached: this is revealed by artificial selection for increase of the terminal interruption of the fourth vein which is, in addition to a decrease of bristles on various sites and a terminal fifth vein interruption, caused by the *Hairless* mutant.

Vein Length in *Hairless*

Hairless is a recessive lethal with dominant morphological effects. The fourth and fifth vein interruptions show incomplete penetrance, i.e. are only present in part of the flies.

Artificial selection (Scharloo 1966) for flies with the shortest fourth vein's, i.e. larger vein interruptions, progressed rapidly (Fig. 9). The penetrance becomes

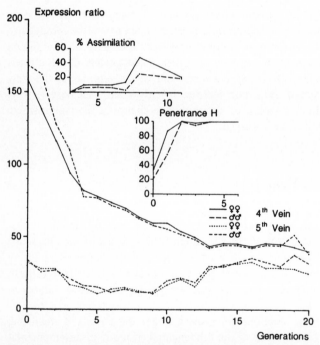

Fig. 9. Selection for a short fourth vein in a *Hairless* stock. Mean relative length of the fourth vein (*upper lines*) and fifth vein (*lower lines*) in flies with a *Hairless* (*H*) mutant. Because *Hairless* is homozygous lethal all individuals selected as parents of the next generation are heterozygous *H/+* In the base population not all *H/+* flies showed a fourth vein interruption but penetrance increased rapidly to 100% by the selection for a shorter fourth vein. After four generations *+/+* flies appear (recognizable because they possess the bristles not present in flies heterozygous for *Hairless*) which have fourth vein interruptions. So selection accumulated modifiers which can cause a fourth vein interruption without the presence of *Hairless* (assimilation). The expression ratio is here the sum of the relative length of the fourth veins of both wings of a fly. The base population was a Pacific cage population in which the mutant $H^{5\,1g}$ was introduced

100% after a few generations of selection and the vein interruptions become larger. In this first stage of the selection variability is large. In later generations when approaching the second cross vein, progress slows down and variability decreases strongly (Fig. 10).

When shifting vein length by temperature change through the same range, similar patterns of change in vein length are observed (Fig. 11). Change per degree Centigrade is large when the vein is near completion and is smaller in regions nearer to the second cross vein. That it is really the range of values, i.e. the position within the wing-vein pattern, is shown in temperature experiments using selection lines in different stages of selection.

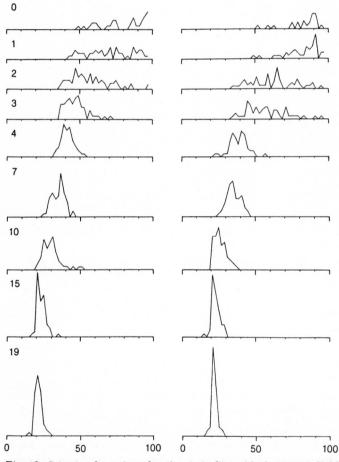

Fig. 10. Selection for a short fourth vein in flies with the mutant Hairless. Frequency distributions of the selection are depicted in Fig. 9. Females *left;* males *right.* In the first three generations part of the *H/+* flies have no fourth vein interruption, only flies with such a vein interruption are represented in the frequency distributions. *Horizontal axis:* Relative length of the fourth vein in one wing

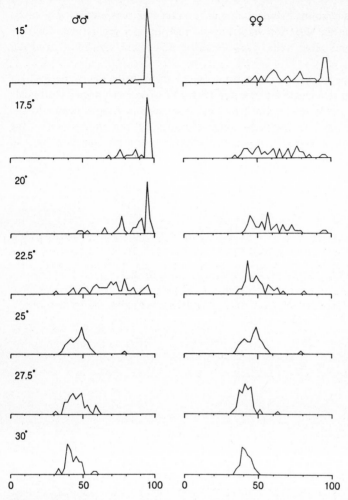

Fig. 11. The effect of temperature on generation five of the selection line depicted in Figs. 9 and 10. Here, the flies with a complete fourth vein (approximately at a relative fourth vein length of 95) are represented

Introduction of second and third chromosomes with dominant markers *Sb* and *Pm* from an unselected stock confirms that their effect in increasing vein length in the selection line depends on the range of vein lengths in which they are acting. When this occurs in the neighbourhood of the second cross vein, their effect is small; when approaching wild-type, the effects become larger (Table 1).

In the *Hairless* case we have full evidence that the same mapping function is acting in the change of vein length by environmental factors, by artificial selection, by chromosome substitution and by internal stochastic factors.

Table 1. The effect of substitution of marked second and third chromosomes of an unselected line into a Hairless selection line selected for a short fourth vein[a]

A	Genotype	sss		sPms	ssSb		sPmSb
	G7	82		123	145		179
	G15	59		64	72		93

B	Substitution		Pm			Sb	
	background	+		Sb	+		Pm
	G7	41	34	63			56
	G15	5		21	13		29

[a] The substitution was performed in two generations of the selection line, generation 7 (G7) and generation 15 (G15) of the selection. In A the mean values of the different genotypes and in B the effect of the substitutions of marked chromosomes in different genetic backgrounds, i.e. when acting in different ranges of the scale. The effects are smaller when they occur at a smaller relative length of the fourth vein. In G15 the effect is smaller than in G7 and the effect of a marked chromosome of the unselected stock is larger when the other marked chromosome is already present and has increased the length of the fourth vein. *sss* three pair of large chromosomes of the selection line; *sPms* introduction of the *Pm* chromosome in the selection line; *ssSb* introduction of one *Sb* chromosome in the selection line; *sPmSb* introduction of one *Pm* and one *Sb* chromosome in the selection line; *Pm* second chromosome of an unselected stock with the dominant eye color mutant *Plum*; *Sb* third chromosome of an unselected stock with the dominant bristle mutant *Stubble*

Correlated Responses

Similar variability patterns were observed in five experiments in which artificial selection was practised for a larger fourth vein interruption of Hairless introduced into different wild-type population. These lines provide an opportunity to study another problem which is relevant to developmental constraints, i.e. genetic correlation. Genetic correlation between two characters is seen as an important cause of constraint. In these experiments we measured, in addition to the length of the fourth vein, the length of the fifth vein. Would there be a correlated responses of the fifth vein when the fourth vein was shortened by artificial selection, and would this correlated response be inevitable when changing the fourth vein?

We found the following patterns of correlated response in three different selection lines:

1. During progress of selection on the fourth vein there was almost no change of the fifth vein.
2. A rapid correlated response of the fifth vein parallel to the response of the fourth vein.

3. In the first stage of the selection a parallel response of the fifth vein, but in the later stages of the selection, while the selection for a shorter fourth vein was still successful, the fifth vein became longer again until it had returned to its original length (see Fig. 10).

The conclusion is obvious: here, genetic correlation is not absolute or inescapable. All things are possible; there are genes which act on both characters in the same direction, there are genes which act specifically on each character separately and there are genes which act in opposite directions on the two characters. This is perhaps a consequence of the fact that the two characters, fourth vein interruption and fifth vein interruption, are new, and had never been submitted to natural selection. The characters are newly created by introduction of the *Hairless* mutant into the wild populations. So natural selection has had no opportunity to generate a fixed relation between the two characters.

Constructional Constraints

Our *Hairless* experiments showed too that constructional features of developmental systems impose constraints on selection response. Selection for shortening of the fifth vein gave an immediate response. The character here was the ratio of the length of the fifth vein (distal to the second cross vein) to the length of the third vein (Fig. 12). When this measurement approaches the zero value, i.e. when there is no fifth vein present behind the cross vein, the response slackens and the frequency distributions show very low variability (Fig. 13). Then a gap appears in the fifth vein proximal to the cross vein, the response takes up again and variability increases.

That genetic change continued when the phenotypic response slackened around the zero value is indicated by two phenomena:

1. The correlated response of the fourth vein continued with the same pace when the fifth vein response halted.

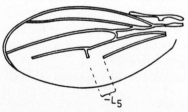

Fig. 12. Measurement of the relative length of the fifth wing vein in *Hairless*. The fifth vein was measured from the second cross vein. When the mean vein length of the fifth vein approached the 0 value, i.e. no vein material left behind the second cross vein, flies appeared with a break proximal to the second cross vein. Then, the length of the break was measured ($- L_5$) and given as a negative value

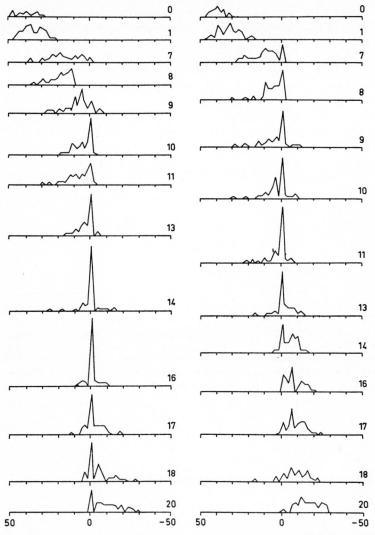

Fig. 13. Selection for a short fifth vein in flies with the mutant *Hairless*. Relative length of the fifth vein measured as indicated in Fig. 11 on the horizontal axis. ♀♀ left, ♂♂ right

2. Because *H* is a recessive lethal, the parents in that generation, in all *H* selection experiments, were always *H/+* and in every generation *+/+* individuals segregated. After a number of generations of selection, in addition to the *H/+* flies with a large fifth vein interruption, *+/+* individuals appeared which showed a small terminal fifth vein interruption. In these individuals the genotype beyond the *Hairless* locus had been changed so much by the selection for modification of the *Hairless* expression on the fifth vein, that they could cause now a fifth vein interruption without the presence of *Hairless*. During the lull in the response of the fifth vein to selection, the increase in frequency of the assimilants and the size increase of the vein interruption continued (Fig. 14). This proves that the change

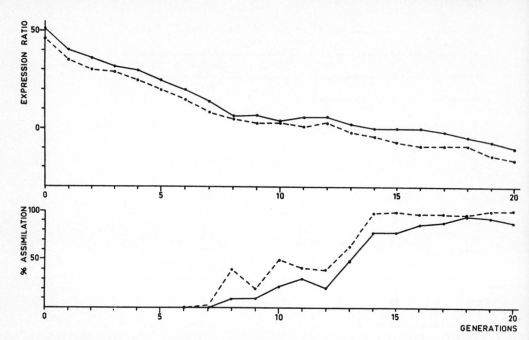

Fig. 14. Selection on the length of the fifth vein in a *Hairless* mutant of *Drosophila*. The expression ratio denotes the relative length of the fifth vein (sum of the two wings of the individual). *Top:* The mean values in the course of generations. *Bottom:* Percent assimilation, i.e. percent of the individuals without *Hairless* segregating in each generation having nevertheless a fifth vein interruption caused by the modifiers accumulated by the selection for a shorter fifth vein in *H/+* individuals

of genotype continued notwithstanding that this was only expressed in a minor way in the change of the phenotypic character selected for.

When the length of the fifth vein was changed by temperature, a similar pattern of changing mean values and frequency distributions was obtained when passing through the range around the zero value as in the selection experiment (Fig. 15).

Moreover, as selection for a shorter fifth.vein was applied to a population obtained from the *+/+* individuals with a fifth vein interruption appearing in the lines where selection was practiced for a larger fifth ven interruption in *H/+* flies, the whole process could be repeated. Also, in this selection line progress diminished and variability decreased when the mean approached the zero value, taking up again after individuals appeared with a break proximal to the second cross vein, concomitant with an increase in variability.

This confirms that the mapping function is to a large extent dependent on the structure of the developmental system which is governed by the presence of the second cross vein.

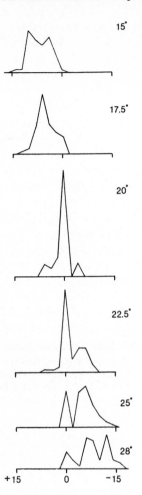

Fig. 15. The effect of temperature on the relative length of the fifth vein in *Hairless* in the line selected 15 generations for a short fifth vein. *Horizontal axis*, the sum of the relative length of the fifth vein in the two wings of a fly

Changing Mapping Functions by Selection

We have already seen that gene-environmental factor/phenotype mapping functions can be changed and are dependent on major mutants, e.g. compare the *Hairless* fourth vein function and the ci^D mapping function and remember that the same mapping function is linear in ci^{D-G}.

Are these mapping functions affected by artificial selection? This question asks (1) Is genetic variation present in populations which affects the shape of the mapping function; (2) Could selection be designed in such a way to use this variation? Schmalhausen (1949) and Waddington (1957) pointed out that stabilizing selection would cause, in addition to the weeding out of disadvantageous mutations, buffering of the development against both environmental and genetic effects. Disruptive selection could be expected to have the opposite effect.

They showed that the effect of temperature on the expression of mutants in *Drosophila* could be changed by artificial selection (Kamshilow 1939; Waddington 1960). However, this was selection specifically against the effect of temperature and the results do not show that here more is involved then this specific effect. There is no proof of a general buffering effect, which should occur when a change of the gene-environmental factor/phenotype mapping function is involved.

That such a specific effect can be involved is shown in the results of stabilizing selection on the length of the fourth vein in the mutant ci^{D-G} in *Drosophila* (Scharloo 1964a, b; Scharloo et al. 1967). In contrast to ci^{D}, where we found a gene phenotype mapping functions with a threshold, this function is linear in its allele ci^{D-G}. Artificial stabilizing selection caused a decrease of genetic variance, environmental variance and, to a smaller extent, of the within-fly variance. However, this did not occur simultaneously as would be expected as change of one underlying developmental process expressed in a gene-environment/phenotype mapping function would be involved. This was different in experiments with disruptive selection on the same character. In lines in which there was disruptive selection with compulsory mating of opposite extremes, the genetic variance, the environmental variance and the within-fly variance increased simultaneously (Scharloo et al. 1967; Scharloo 1970). Temperature experiments and the study of the frequency distributions both in selection and temperature experiments show that the linear GEPM has been changed by the disruptive selection in a threshold-like GEPM. Moreover, the extent of the threshold region could differ between different lines showing that the precise shape of the GEPM is under genetic control (Fig. 16). (Scharloo et al. 1972).

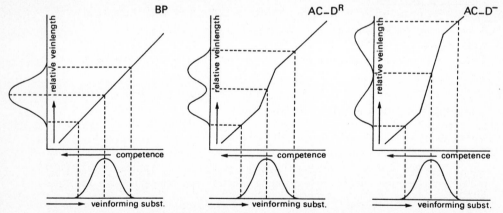

Fig. 16. Model for the development of the fourth vein of ci^{D-G} in the base population and two lines obtained by a system of disruptive selection (anti-canalizing selection) on the relative length of the fourth wing vein. Formation of the fourth vein is supposed to be controlled by: (1) the concentration of a vein-forming substance in the wing varying according to a normal distribution. (2) A gradient of competence to react to the vein-forming substance with the formation of vein material. The competence decreases from wing base to wing tip. The gradient determines the pattern of vein formation and the variability of the relative vein length. The difference of the extent of the threshold zone between the lines is concluded from a difference in the distance between the modes of the bimodal frequency distributions generated by the selection in the two lines

Rendel et al. (1966) showed that stabilizing selection on the number of scutellar bristles in the scute mutant could generate canalization of this number. Their selection involved both selection for a fixed number of two and selection against variation. They found that the width of the bristle class 2 could be increased considerably compared to higher and lower bristle classes.

Rendel (1967) stated that such stabilizing selection could only generate canalization when the character under selection was affected by a major mutant. A situation which he characterized as the presence of an uncontrolled major gene.

However, we succeeded in obtaining canalization of scutellar bristle number at eight based on genetic variation in wild populations. In two wild populations of independent origin artificial selection was applied for increased bristle number.

When bristle number increased variation increased too, instead of the almost constant number of four with only an occasional fly with five or three bristles, the bristle number varied between 4 and 15. When both populations had approximately eight bristles, they were crossed to form the base population and then submitted to artificial stabilizing selection: only flies with eight bristles were selected as parents of the next generation. This caused the following changes (Scharloo et al. 1972; Scharloo 1987, in press):

1. Variability decreased strongly as a consequence of a decrease of genetic variability.
2. The width of the eight bristle class decreased relative to the seven- and nine-bristle classes.
3. The effect of the genetic difference between ♀♀ and ♂♂ on bristle number is halved.

The last two points indicate that the stabilizing selection generated canalization of scutellar bristle number at eight. Further analysis showed that the canalization is not a canalization of total bristle number as shown by Rendel (1959, 1965), but an increase of the precision of development per bristle site as suggested by Robertson (1965). This canalization is caused by changes on all three large chromosomes and is built up gradually during the course of selection.

Constraints by Interaction with the Environment

Until now we described experiments in which the constraints were determined internally, i.e. a consequence of the organization of development. However, constraints can also be a consequence of interaction between properties of the organisms and its environment. This occurred in our selection for increased size of the anal papillae in larvae in *Drosophila melanogaster*. After a few generations larvae were found with anal papillae looking like inflated balloons, which were in cross-section much larger than the diameter of the rest of the larvae. The surface of the papillae appeared rather delicate. Moreover, it interfered with the movement of larvae in the solid medium. This probably caused the high mortality of the larvae with ballooning papillae and constrained selection response (te Velde 1985; te Velde et al., in preparation).

It is tempting to think that this constraint could be removed by performing selection on a liquid medium.

Limitations of the Quantitative Genetic Approach

We have seen that developmental constraints are revealed in quantitative genetics. Do these phenomena support the theories of the punctuationalists that developmental constraints are a consequence of the intrinsic organization of developmental systems and that these constraints could explain stasis, i.e. the long periods of constancy of species in evolution?

We have shown that in the experiments performed in quantitative genetics, the presence of constraints is affected by genetic variation and can be shaped by stabilizing and disruptive selection. Constraints are affected both by major mutants and by polygenic variation. Developmental constraints can be built up gradually by selection. Then, of course, developmental constraints built up by selection can influence further effects of selection.

Available evidence does not seem to support the theories implying that developmental constraints are based on fundamental properties of developmental systems (see e.g. Kauffman 1983; Alberch 1982). Constraints as they were revealed in quantitative genetics are subject to genetic variaition and can be built up by selection.

The question arises now whether the results of quantitative genetics are a consequence of the very nature of these experiments. The characters of quantitative genetics are abstractions; mere numbers represent often a complex phenotypic character. Are the results of these experiments to a large extent a consequence of the definition of the characters and the design of the selection regimes applied?

For instance in the experiments on the scutellar bristles of Rendel and Scharloo, selection was on numbers with only passing reference to the fixed pattern in which the bristles appear.

Would more fundamental features of development be revealed when selection is applied on the pattern aspects of the character? Would then perhaps fundamental aspects of the organization of the developmental system in the thorax constrain selection response? Constraints which had to do with the formation of the rectangular pattern of scutellar or dorsocentral bristles within the mesothorax or even with the fundamental properties of segmental development?

Selection on a Pattern

Of course, one could argue that also these aspects had their evolutionary history under the action of natural selection. However, these properties would have their origin far earlier in the origin of the species, and would concern far more fundamental developmental features than mere bristle number. Therefore, we decided to start artificial selection on the pattern of dorsocentral bristles in *Drosophila*. There are four such bristles in a rectangular pattern on the mesonotum. We tried to shift the position of the scutellar bristles relative to the anterior and posterior borders of the mesonotum.

We had some expectations for the result. There are theories on the formation of bristle patterns (see Meinhardt 1982). In his models, the formation of bristle patterns depends on: (1) An autocatalytic bristle-forming substance which generates (2) The production of an inhibiting substance responsible for inhibition of the formation of other bristles in the neighbourhood. Now in terms of this model one would expect that it would be difficult to shift the anterior bristle in a posterior direction because then its position would approach the site of the posterior bristle. Then, mutual inhibition would perhaps prevent selection response.

Or could it be related to the more fundamental level of segmentation or compartmentalization? Could compartment borders, which are between scutum and scutellum (Garcia-Bellido et al. 1976) and are supposed to be present in front of the dorsocentral bristles on the border between scutum and presentum, inhibit a shift of the anterior dorsocentral bristles to a more anterior position?

In fact, progeny tests and the comparison of the variance in long inbred lines with the variance in the base population revealed a heritability of approximately 40%. Artificial selection made for the position of the anterior bristles had an immediate result. However, the selection response was extremely asymmetric: the selection for a more anterior position made no progress at all, while there was a considerable response in the line selected for a shift of the anterior dorsocentral bristle to a more posterior position. The posterior bristle is then shifted posteriorly, too. This is, perhaps, a consequence of an inhibiting effect of the anterior bristle. The pattern of the four scutellar bristles shifted simultaneously in a posterior direction.

What the precise interpretation of these results will be awaits further analysis. However, it seems that quite fundamental properties of the developmental system of bristle patterns and segmentation are involved. This kind of experiment seems to open up exciting possibilities for the exploration of the role of properties of developmental organization in evolution (see Scharloo 1987, in press).

Conclusions

Constraints in development which affect the expression of genetic variation are revealed in quantitative genetic studies. They can be shown by analysis of the patterns of response caused by artificial selection and of the variation patterns during that response. The conclusions can be supported by chromosome exchange between unselected stocks and lines at different stages of the selection and by studying the effect of environmental factors. Genetic differences and environmental factors are affecting quantitative characters often via the same developmental pathways. This is revealed by the shape of gene-environmental factor/phenotype mapping functions (GEPMs). The shape of the GEPMs is affected by mutations and by artificial stabilizing and disruptive selection. Thus, the constraints are not an inevitable consequence of the structure of developmental processes. In these cases the constraints are a consequence of the selection history of the character, thereby affecting the response to further selection. Constructional features involved in the development of quantitative characters could play a more fundamental role.

It is argued that the very nature of the experiments in quanitative genetics could preclude the demonstration of the more fundamental constraints. It is suggested that selection on pattern aspects of morphological characters is suited to reveal such fundamental constraints.

It is now the time to incorporate the new insight in the development of morphological patterns which was obtained in the last 15 years in *Drosophila* (see e.g. Malacinsky and Bryant 1984) into evolutionary theory. We must try to design experiments in which change by artificial selection reveals what effect developmental organization has on the possibility of change.

References

Alberch P (1980) Ontogenesis and morphological diversification. Am Zool 20:653–667

Alberch P (1982) Developmental constraints in evolutionary processes. In: Bonner JT (ed) Evolution and Development. Springer, Berlin Heidelberg New York, pp 313–332

Clayton GA, Robertson A (1957) An experimental check on quantitative genetical theory. II. The long-term effects of selection. J Genet 55:152–170

Dunn RB, Fraser AS (1959) Selection for an invariant character, vibrissae number, in the house mouse. Aust J Biol Sci 21:506–523

Falconer DS (1981) Introduction to quantitative genetics, 2nd edn. Longman, London

Fraser AS, Kindred BM (1960) Selection for an invariant character, vibrissae number in the house mouse. II. Limits to variability. Aust J Biol Sci 13:48–58

Garcia-Bellido A, Ripoll P, Morata G (1976) Developmental compartmentalization in the dorsal metathoracic disc of *Drosophila*. Dev Biol 48:132–147

Kamshilow MM (1939) Selection as a factor alternating the dependence of a character on variations of environmental conditions. Comp Rend Acad Sci l'URSS 23:362–365

Kauffman SA (1983) Developmental constraints: internal factors in evolution. In: Goodwin BC, Holder N, Wylie CC (eds) Development and evolution. Sixth Symposium of the British Society for Developmental Biology. Cambridge University Press, Cambridge, pp 195–225

Malacinsky GM, Bryant SV (1984) Pattern formation. Collier Mac Millan, New York

Maynard Smith J, Sohndi KC (1960) The genetics of a pattern. Genetics 45:1039–1050

Meinhardt H (1982) Models of biological pattern formation. Academic Press, London

Rendel JM (1959) Canalization of the scute phenotype of *Drosophila*. Evolution 13:425–439

Rendel JM (1965) Bristle pattern in scute stocks of *Drosophila melanogaster*. Am Nat 99:25–32

Rendel JM (1967) Canalization and gene control. Logos, London

Rendel JM, Sheldon BL, Finlay DL (1966) Selection for canalization of the scute phenotype. II. Am Nat 100:13–31

Robertson A (1965) Variation in scutellar bristle number. An alternative hypothesis. Am Nat 99:19–24

Scharloo W (1962) The influence of selection and temperature on a mutant character (ci^D) in *Drosophila melanogaster*. Arch Neerl Zool 14:431–512

Scharloo W (1964a) The effect of disuptive and stabilizing selection on the expression of a cubitus interruptus mutant in *Drosophila*. Genetics 50:553–562

Scharloo W (1964b) Mutant expression and canalization. Nature 203:1095–1096

Scharloo W (1966) Pattern formation and canalization. Genen Phaenen 11:1–15

Scharloo W (1970) Stabilizing and disruptive selection on a mutant character in *Drosophila melanogaster*. III. Polymorphism caused by a developmental switch mechanism. Genetics 65:693–705

Scharloo W, Hoogmoed HS, ter Kuile A (1967) Stabilizing and disruptive selection on a mutant character in *Drosophila*. I. The phenotypic variance and its components. Genetics 56:709–726

Scharloo W, Zweep A, Schuitema KA, Wijnstra JC (1972) Stabilizing and disruptive selection on a mutant character in *Drosophila*. IV. Selection on sensitivity to temperature. Genetics 71:551–566

Schmalhausen II (1949) Factors of evolution. Blakiston, Philadelphia

te Velde J (1985) The significance of the anal papillae in salt adaptation of *Drosophila melanogaster*. Thesis, University of Utrecht

te Velde JH, de Ruiter BLA, Scharloo W (1987) Selection for large and small anal papillae in larvae of *Drosophila melanogaster* on media with different osmotic pressure (in prep)

Waddington CH (1957) The strategy of the genes. Allen & Unwin, London

Waddington CH (1960) Experiments on canalizing selection. Genet Res 1:140–150

Yoo BH (1980) Long-term selection in a quantitative character in large replicate populations of *Drosophila melanogaster*. Genet Res 35:1–17

Chapter 8 Nonrandom Patterns of Mutation are Reflected in Evolutionary Divergence and May Cause Some of the Unusual Patterns Observed in Sequences

G. B. GOLDING[1]

Introduction

Very little is known, as yet, about spontaneous mutation and even less is known about how these random changes are translated into a useful and adapted organism through the action of selection. Both of these processes are complicated by the significant effects which even small mutation rates and small selection coefficients can have over long periods of time. These processes will be the subject of this chapter and although only some fragments of information are beginning to become apparent, many of them are contrary to common expectations.

It is common practice in many scientific studies to bestow upon mutations and substitutions, a regularity and consistency that is normally attributed to radioactive decay. But recent experimental studies of spontaneous and induced mutations are demonstrating that this is not a realistic assumption. These studies invariably find that mutations are affected by a large number of extrinsic and intrinsic factors (Drake 1970) and that they do not act (as a collection) with the regularity of a physical process.

An analysis of evolutionary changes has also demonstrated that observed substitutions reflect many of the unusual patterns that are shown by spontaneous mutations. Among the first papers to notice this effect were those which documented the effect of an excess of transitions compared to transversions. The suggestion that transition mutations (a mutation of an A:T pair to or from a G:C pair) should be more frequent than transversion mutations (all others) has been clear from the initial studies of a molecular basis for mutation (Watson and Crick 1953). Experimental demonstrations that such a pattern in mutations is commonly observed are reviewed by Drake (1970; p. 178). But experimental and theoretical evidence that transitions occur more frequently than transversions does not necessarily mean that a similar pattern would be observed in evolutionary changes. The proof for this was provided by Vogel and Kopun (1977), who noted that transitions were more common among substitutions in hemoglobins. The effects of a transition bias on estimates of sequence divergence have been well analyzed by Kimura (1980, 1981), Takahata and Kimura (1981), Aoki et al. (1981) and Gojobori et al. (1982). In general, the major effect of a strong transition bias is an increase in the number of sites with more than one subsitution.

[1] Department of Biology, York University, 4700 Keele Street, North York, Ontario, M3J 1P3

Genetic Constraints on Adaptive Evolution
Ed. by V. Loeschcke
© Springer-Verlag Berlin Heidelberg 1987

The recent experimental data is, however, providing evidence for more kinds of biases and a greater prevalence of nonrandomness in spontaneous mutations. Many of these studies suggest that rather unusual types of mutations and complex biases can be induced or observed. Some of these patterns will be reviewed here and evidence will be presented that these same types of unusual patterns can also be observed within evolutionary substitutions. The generation of genetic variation is a necessary prerequisite of evolution and how these spontaneous mutations are affected by natural selection must be an important constituent of the evolutionary process. Whether these mutations form a constraint that selection must act against, or whether they provide a greater plasticity and eliminate some of the constraints of natural selection will be discussed.

Expectations

The first clearly understood molecular basis for mutation was that due to base analogues. These are mutagens which have a chemical structure similar to a standard base but sufficiently different that they show a greater frequency of mispairing. A mutation results when they pair with an incorrect base during DNA replication.

Another class of mutations that are moderately understood are those caused by chemical mutagens. These include agents such as ethylmethane sulphonate (EMS), hydroxylamine and nitrous acid. For example, the major effect of the latter mutagen is to deaminate cytosine or adenine such that these bases will then incorrectly pair with adenine and cytosine rather than guanine and thymine respectively. This causes C:G to T:A or A:T to G:C transition mutations.

The final class of mutations that have been extensively studied are those caused by high energy forms of radiation. Radiation can act like a chemical mutagen and can cause chemical alterations to individual bases. However, irradiation has a more important effect in inducing various host repair functions to rectify more extensive damage such as chromosome breaks.

For many who do are not actively involved in the study of mutagenesis these mechanisms, with the exception of the last effect of radiation, form the basis of their conceptions of mutation. That is, there are molecular lesions which affect a single nucleotide site and these sites are then either repaired or misrepaired and result in a mutation. One of the key themes behind such an impression is that mutations affect nucleotide sites in isolation.

The induction of repair processes by radiation damage is one of the mechanisms which suggests that mutation is not a simple process acting in isolation from the rest of the DNA. The repair processes that are invoked in *E. coli* are highly adapted, complex systems. A large number of genes have been isolated which are involved in or can influence the repair of DNA damage (e.g. Kornberg 1980, p. 620). These repair processes are sufficiently complex that there are different pathways specialized for different types of lesions. For example, uracil N-glycosylase is an enzyme specialized to remove uracil from DNA, this leaves a site missing a base which is then further repaired. Another pathway is termed SOS repair and is an extreme mechanism to repair otherwise lethal types of DNA lesions. The induction of SOS repair typically leads to an enhanced mutation rate, perhaps due to reduced fidelity of the polymerase. This pathway and the induction of other lesions by radiation have been used to show that some repair tracts can be relatively long, up to 40—50 bp (Snyder and Regan 1982). If this repair is done with an error-prone or a less efficient polymerase there may be several mutations which occur as a result of a single lesion. Indeed, the original lesion might be correctly repaired but another mutation might be caused elsewhere within the repair tract.

Thus, the first step necessary for evolution, the generation of genetic variability, can be a complex process. The induction of different repair pathways could generate biases in the patterns of mutation that would be very difficult to comprehend. Besides the possibility of different mutational biases for different mechanisms and the possibility of multiple mutations from well-known mechanisms of repair, there is also the possibility that neighbouring bases influence the patterns of mutation.

Nearby Nucleotides can Influence the Type and Frequency of Mutations

Some of the possible ways in which nearby nucleotides can influence mutation are relatively obvious. For example, the presence of many nearby A:T base pairs might allow for a greater degree of natural unwinding of the DNA helix and may expose the nucleotides to the action of chemicals which would otherwise not be able to penetrate the helix.

This kind of an influence would not only affect the rate of mutation at a particular site but it could also affect the type of mutational change. Further, this kind of influence could potentially extend over a great distance. Although this particular influence has not been proven with experimental data, there are other influences of the neighbouring DNA that have been detected. One of the first studies to detect an influence of neighbouring bases was a study by Koch (1971) in T4. He analyzed 2-aminopurine induced mutations of UAG and UAA nonsense codons to UGG or UGA codons in the *rII* gene of T4. In each case the nonsense mutant was located at the same site and an A:T to G:C mutation was induced. The only difference between the two nonsense codons is the presence of a G or an A in the third codon position. But the rate at which UAG mutated to UGG was found to be 23 times higher than the rate at which UAA mutated to UGA. The difference in the rate of mutation at the second codon position can only be attributed to an unknown influence of the base present at the third codon since the remainder of the genomes were identical. That this effect extends beyond the adjacent base was also confirmed as it was found that the rate at which UAG mutated to CAG was 4.2 times higher than the rate at which UAA mutated to CAA. Again, the only difference in the genomes which could affect the different rates of mutation at the first position being a single nucleotide difference in the third position.

Other studies since this time have confirmed a host of possible effects due to neighbouring base pairs. The distances over which nearby bases can exert an effect may also be extensive since Conkling et al. (1980) were able to demonstrate differences in reversion frequencies dependent on changes in a single nucleotide up to 30 bp distant. The influence of nearby nucleotides may be due to secondary causes involved with DNA metabolism rather than directly with mutagenesis or repair. For example, a local or transient depletion of nucleotide precursors may enhance the probability of a mutation. The data of Fersht (1979) and Kunkel et al. (1981) show that the mutation rate at a specific site depends on the rate of incorporation of the next nucleotide in the sequence. Evidence that mutation depends on the correct balance of nucleotide precursors is reviewed by Kunz and Haynes (1981) and Kunz (1982).

All of this would have no effect on evolutionary studies if these processes were not reflected in substitutions between species. However, there is another example of neighbouring bases affecting mutation which was originally discovered through comparative studies of sequences from many species. Josse et al. (1961) noted that vertebrate calf thymus DNA had a deficiency of CG doublets which was not present in bacterial DNA. A bias against CG pairs has since been confirmed across many different vertebrate taxa (McClelland and Ivarie 1982). These CG doublets are thought to form a hot spot for mutation. The mutation of the cytosine nucleotide being enhanced by the presence of the adjacent guanine well beyond the level expected when other nucleotides are immediately adjacent. Salser (1977) suggested this higher mutability was due to the methylation of the 5′ carbon of cytosine. Methylation is one of the common modifications of nucleotides and, in vertebrates, a cytosine that is a member of a CG doublet is often methylated. But this alteration of the cytosine also permits a spontaneous deamination forming a thymine nucleotide (Bird 1980).

The specific patterns of methylation are thought to be involved in the control of gene expression. Because of this there may be some genes that require the presence of many CG pairs. These pairs must then be constantly regenerated because they form mutational hot spots and are continually being changed into TG pairs by mutation. It has been suggested that the alpha globin genes require an abundance of methylated CG pairs for their correct expression and Smith et al. (1985) presented evidence for the continual generation of CG pairs in these genes. Similarly, in the alpha interferon genes of man, when a G mutates to a C it may be preferentially followed by another G (Golding and Glickman 1986). This preference creates a CG dinucleotide and again demonstrates the influence of an adjacent base. Independent of the actual details, which will probably change for each gene, the theme is clear. There can be preferential mutation of some nucleotides dependent on the type of nucleotides nearby and this places constraints on natural selection. There can be pairs which are selectively maintained (as above) and there can be pairs which are selectively deleterious due to their increased rate of mutation.

Repeats can Template Mutations

Other kinds of biases may be present in mutational events and one of these may be an influence of the repetitive structure of the local DNA. A cause and effect relationship between direct repeats and the presence of frameshifts was first proposed by Streisinger et al. (1966). They noted that frameshifts were often located immediately adjacent to a direct repeat. They proposed that the direct repeat permitted slippage by the DNA polymerase during replication. The slippage led to incorrect copying of the template and explains the coincidence of these two phenomena.

The mechanism can also explain more extensive deletions (Albertini et al. 1982). One possible way in which this might occur is diagrammed in Fig. 1A. Here, a pair of repeats are located near each other but are not adjacent. During replication of the first repeat the template strand slips out of register. This misalignment is stabilized by the base pairing between the two repeats. Replication continues past the repeat using the misaligned template. The result is a deletion of one copy of the repeat and all of the intervening sequence. There is good evidence that at least some deletions are created by this mechanism and indeed, there are several clinical human hemoglobin diseases caused by deletions that are generated in this way (Harris 1980).

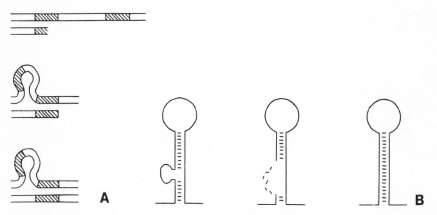

Fig. 1. A A model to explain the presence of deletions associated with direct repeats as proposed by Streisinger et al. (1966). During DNA replication of two direct repeats (*cross-hatched areas*) synthesis proceeds as far as the first repeat. At this time, the DNA misaligns, the misalignment is stabilized by the correct base pairing between repeats and synthesis proceeds. The result is the deletion of one copy of the repeat and all of the intervening sequence (After Harris 1980). **B** A model to explain deletions associated with inverted repeats or palindromes. Besides being able to form a hydrogen bond with a complementary strand of DNA, intrastrand hydrogen bonds can also be formed. This is an example of such a palindrome which leaves several bases unpaired. The single-stranded region may be recognized and attacked by specific enzymes (such as S1 nuclease, whose function is to do just this) and then repaired by a DNA polymerase. Again in this case, the DNA polymerase may correctly copy a template, but may use the template provided by the complementary strand of the palindrome rather than the complementary strand of the DNA helix. In this case the palindrome is perfected and the previously unpaired region is deleted (After Ripley 1982)

Inverted repeats or palindromes may also be involved in the generation of some deletions/insertions as shown in Fig. 1B (Ripley 1982). These are regions of the DNA where correct hydrogen bonding can occur between the bases on a single DNA strand. This figure shows an imperfect palindrome with a small region of unpaired bases. A lesion is made in the DNA near the palindrome (the cause of the lesion is not central to this argument and it may even be associated with the presence of the palindrome). When repair occurs, the opposite strand of the palindrome is used as a template. This template is stabilized by the base pairing between opposite strands of the inverted repeat. The result is a new sequence with the previously unpaired region deleted.

Both of these mechanisms are able to generate frameshift and deletion mutations. But with a little modification both mechanisms are also able to generate base substitutions. This is shown in Fig. 2. A possible role for direct repeats in the generation of base substitutions is shown in Fig. 2A. This is a modification of the method suggested by Streisinger et al. (1966) in Fig. 1A. The method begins with two imperfect repeats which differ by several nucleotides. During replication a similar misalignment occurs but after synthesis of the direct repeat, the two strands realign and synthesis then proceeds normally. Instead of generating a deletion, this produces a perfect copy of the direct repeat with concomitant base substitutions.

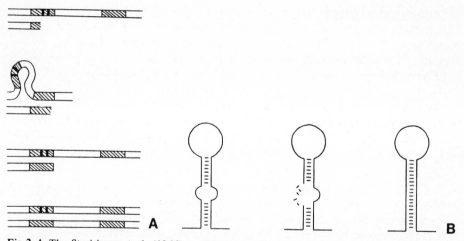

Fig. 2. A The Streisinger et al. (1966) model can be adapted to explain the presence of base substitutions associated with direct repeats. Consider two imperfect direct repeats. Again a misalignment can occur during replication (as in Fig. 1A). Synthesis proceeds using the second copy of the repeat as a template and then the misalignment is corrected. After a correct alignment, DNA synthesis continues and the second repeat is copied again. The result would be base substitutions which perfect the direct repeats. This is only one possible mechanism and not the sole mechanism which could explain the association of mutations with direct repeats. **B** Similarly, the presence of mutations associated with palindromes can be explained if the unpaired regions were of the same length. In this case when a gap is repaired, the palindrome would again be perfected but the result would be interpreted as base substitutions rather than deletions. It should be apparent that mixtures of base substitutions and deletions/insertions might also occur when mutations are templated by either direct repeats or palindromes

Similarly, base substitutions can result if a polymerase is using the complementary strand of a palindrome as a template. This is shown in Fig. 2B. Again, the inverted repeat is perfected and the originally unpaired bases are replaced by bases which can pair and would be observed as mutations.

There are other mechanisms by which these repeats could be involved in the production of mutations. For example, a simple conversion or correction of one repeat upon the other would generate the same result. In addition to the creation of deletions/insertions and the creation of base substitutions, both of these mechanisms could generate mixtures of the two types of mutations ("complex" mutations) as the result of a single event.

The action of a mechanism involving direct repeats has been observed to occur in vitro (Kunkel and Alexander 1985). It has also been inferred in *E. Coli,* T4 and yeast (Ripley 1982; Ripley and Glickman 1983; Drake et al. 1983; de Boer and Ripley 1984). However, this does not imply that this is a mechanism leading to substitutions which are important in evolution or that such a mechanism acts in higher eukaryotes where the DNA is tightly bound with histones.

Substitutions are also Influenced by the Repetitive Structure of DNA

In order to determine if such a mechanism might be important in evolution, I have analyzed substitutions in evolution and present some of these results here. The sequences examined include intraspecific changes within *Drosophila melanogaster Adh* genes (Kreitman 1983), interspecific substitutions within *Drosophila melanogaster, D. simulans* and *D. mauritiana Adh* genes (Cohn et al. 1984), the chorion genes of *Bombyx mori* (Rodakis et al. 1984), the actin genes of *Strongylocentrotus franciscanus* (Foran et al. 1985) and the gamma globin genes of man (Slightom et al. 1980). The sequences and their characteristics are given in Table 1. The intraspecific changes noted by Kreitman (1983) were associated with a fast – slow allozyme polymorphism at the *Adh* locus. For the sake of convenience these two groups have been analyzed separately. The substitutions in these genes will demonstrate that repeats have also been a potential factor in evolutionary divergence.

These genes were chosen because their sequences are known to be very closely related. This high degree of homology permits at least limited inference of the substitutions creating their divergence and of their ancestral sequence structure. A computer program was used to search for any direct or inverted repeats that may have templated the substitutions in these genes as per Fig. 2 (Golding and Glickman 1986).

Some repeats are expected to occur within a DNA sequence by chance alone. This is because DNA consists of only a four-symbol code and eventually all combinations of bases must be repeated. The repeats were analyzed for significance as a function of their length and their proximity to the substitution. The probability of finding a continuous repeat by chance increases as the length of the repeat decreases or as the distance between them increases. The latter indicating that a perfect repeat will eventually be found if a sufficiently long enough stretch of DNA is searched.

Table 1. Source and characteristics of some closely related DNA sequences

Sequences	Species	Number of substitutions[a]	Base of Composition				Length[b]	Reference
			T	C	A	G		
Adh Fast variants	Drosophila melanogaster	27	0.25	0.23	0.31	0.22	2796	Kreitman (1983)
Adh Slow variants	Drosophila melanogaster	33	0.25	0.23	0.31	0.22	2723	Kreitman (1983)
Adh[c]	Drosophila melanogaster Drosophila mauritiana Drosophila simulans	65	0.21	0.29	0.27	0.23	831	Cohn et al. (1984)
Chorion[e]	Bombyx mori	65	0.32	0.17	0.26	0.25	1388	Rodakis et al. (1984)
Gamma globin[d]	Homo sapiens	64	0.26	0.20	0.26	0.27	1483	Slightom et al. (1980)
Actin[e]	Strongylocentrus franciscanus	76	0.26	0.26	0.26	0.22	1546	Foran et al. (1985)

a Each insertion/deletion is counted as one substitution regardless of length
b Includes positions where insertions/deletions have occurred
c Pairwise comparisons
d Pairwise comparisons between Gγ and two allelic Aγ
e Two members of a multigene family

The probability of a particular sequence of DNA is given by the product of the component base frequencies. This is the probability of an adjacent repeat assuming that the immediately adjacent DNA is of a random origin. The probability of finding a perfect repeat is then assumed to increase as more and more DNA is searched according to a geometric (or waiting time) distribution. This provides a metric, or a measure of the probability. Since this metric depends on the total length of the sequence being searched, the number and density of other substitutions, the number of uncertain bases in the sequence and other factors, simulations were performed to allow for these factors. The metric was used with the simulations to provide empirical levels of significance. For each gene being examined, a simulation was conducted that placed the same number of substitutions at random on the observed sequence. This was repeated ten times for each gene. For example, there are 65 mutations in the chorion genes (Table 1) spread over 1234 bp. In the simulations, ten replicates of these sequences were created (each exactly as observed) and on each sequence 65 random mutations were placed. These were then analyzed in exactly the same manner as the actual observed mutations. A total of 650 simulated substitutions were examined and compared with the observed substitutions. A 5% level of significance, which is specific for the chorion genes, can be found from the 650 mutations.

In general, these simulations (and others) lead to very stringent levels for a repeat to be significant. Often, even a direct repeat of 5 immediately adjacent would not be statistically significant, and yet it would commonly be given biological importance. For this reason, the sequences were also searched for less extensive repeats capable of templating the substitutions. These repeats had to consist of at least 6 bp within a distance for 50 bp. The search for statistically significant repeats did not extent more than 100 bp beyond the substitution. Both of these distances are probably a conservative estimate of the length over which such interactions can occur. Gene conversions and deletions are known between repeats separated by several thousands of base pairs (Müller-Hill and Kania 1974; Harris 1980).

The number of substitutions which have potential repeats capable of templating the observed substitutions are shown in Table 2. As can be seen, the vast majority

Table 2. Substitutions with nearby repeats[a]

Gene	Nearby direct repeats	Nearby palindromes	Both repeats nearby	Total potentially templated	Total not templated
Adh Fast	7 (26%)	4 (15%)	5 (19%)	16 (59%)	11 (41%)
Adh Slow	3 (9%)	3 (9%)	5 (15%)	11 (33%)	22 (67%)
Adh[b]	11 (17%)	10 (15%)	9 (14%)	30 (46%)	35 (54%)
Chorion	10 (15%)	13 (20%)	5 (8%)	28 (43%)	37 (57%)
Gamma Globin[b]	18 (28%)	8 (13%)	3 (5%)	29 (45%)	35 (55%)
Actin	13 (17%)	15 (20%)	10 (13%)	38 (50%)	38 (50%)

[a] Repeats must include at least six continuous nucleotides, must template the observed substitution and must be within 50 nucleotides distant

[b] Pairwise comparisons

Table 3. Number of base substitutions[a] with significant nearby direct repeats and palindromes

Gene	No. with direct repeats	No. with palindromes	Total No. of substitutions
Adh Fast	1 (4.2%)	2 (8.3%)	24
Adh Slow	0 (0.0%)	2 (6.7%)	30
Adh[b]	9 (16.4%) ***	7 (12.7%) **	55
Chorion	7 (12.1%) **	3 (5.2%)	58
Gamma Globin [b]	2 (3.5%)	2 (3.5%)	57
Actin	1 (2.0%)	5 (9.8%)	51
Total	20 (7.3%) *	21 (7.6%)*	275

[a] Excluding insertions and deletions

[b] Pairwise comparisons which are not necessarily independent

Probabilities are calculated assuming normality

* $P < 0.05$, ** $P < 0.01$, *** $P < 0.001$

of substitutions have nearby repeats that are capable of templating their occurrence. Less than 70% of the mutations do not have a direct or inverted repeat within close proximity of the substitution and many of the substitutions have both types of repeats nearby.

To determine whether there is an excess of significant repeats, the number of extensive, nearby repeats were compared with the expectations provided by the simulations. These results are presented in Table 3. Again, the probability level below which 5% of substitutions should fall is determined from the simulations, but table 3 shows that many of the genes have more than 5% of the substitutions with repeats above this probability level. Many of the substitutions have very extensive repeats and these exist in significant excess of the number expected for four of the twelve comparisons. Overall, there is again a significant excess, indicating that a least some of the substitutions many have been created via the influence of DNA repeats.

Examples of some of the direct repeats illustrate several of the unusual features that can result. An example of a substitution that perfects an immediately adjacent direct repeat is shown in Fig. 3. The mutation is a G to T transversion and creates a 6 bp direct repeat. A greater proportion of the repeats are not immediately adjacent. Some repeats, such as those illustrated in Fig. 4 potentially cause two substitutions as a result of a single event.

<u>Adh</u> Consensus   GCT⌐GCCGTC⌐GCCGGCTGA
                                            *
<u>Adh</u> Fast Variant   GCT⌐GCCGTC⌐GCCGTC⌐TGA

Fig. 3. Direct repeats in intron 1 of the *Adh* gene of *Drosophila melanogaster*. These repeats include six base pairs with the repeats immediately adjacent. Within these direct repeats there has been a G to T transversion in one of the *Adh* variants that could have been templated by these direct repeats (sequence data from Kreitman 1983)

Often evolutionists are more concerned with coding sequence than with flanking or intervening sequences. It would be desirable to have some indication that these mechanisms also work in sequences which code for a protein. A significant excess of repeats have been found elsewhere, within the coding sequence of interferon genes (Golding and Glickman 1986). Here, Fig. 5 shows a direct repeat consisting of 8 bp within 7 bp that templates a silent substitution in an isoleucine codon of *Adh*. Figure 6 illustrates direct repeats which potentially template a base substitution which causes an amino acid replacement. In this case, 8 bp repeats within 16 bp, that may have been responsible for the replacement of a glutamine in *D. melanogaster* by a lysine residue in *D. simulans*. There are only four other replacements among the 179 residues sequenced by Cohn et al. (1984) and these may have been templated by less extensive repeats (Table 2).

The direct repeats in Fig. 7 are an example of a potential template which would cause a multiple substitution including one silent substitution and the replacement

D. melanogaster
       TCCATGCAGCGATGGAGGTTA
             * *
D. mauritiana
       TCCATGGAGGGATGGAGGTTA

Fig. 4. Direct repeats in the introns of *Adh* genes. These repeats include seven base pairs separated by a single base. The repeats template a C to G transversion and a second C to G transversion only two base pairs away. Thus, these two substitutions may be the result of a single event (sequence data from Cohn et al. 1984)

D. melanogaster
       TGCCGGTCTGGGAGGCATTGGTCTGGACA
                            *
D. simulans
       TGCCGGTCTGGGAGGCATCGGTCTGGACA

Fig. 5. Direct repeats in the coding region of the *Adh* gene. In this example, an A:T – G:C transition perfects an 8 direct repeat. The repeats are separated by 7. The mutation results in a silent substitution in the third position of an isoleucine codon (sequence data from Cohn et al. 1984)

D. melanogaster
       CCACCAAGCTGCTGAAGACCATCTTCGCCCAGCTGAAG
                                    *
D. simulans
       CCACCAAGCTGCTGAAGACCATCTTCGCCAAGCTGAAG

Fig. 6. Direct repeats in the coding region of *Adh* which template an amino acid replacement. These direct repeats consist of 8 separated by 16 and potentially template a C to A transversion. This substitution is responsible for a glutamine-lysine replacement (sequence data from Cohn et al. 1984)

Chorion HC-B.13
       GTAGAGGTTGTGGCTGCGGTTGTGGAGGTTGCGGCTCTAG
                                    *   *
Chorion HC-B.12
       GTAGAGGTTGTGGCTGCGGTTGTGGAGGTTGTGGCTGTAG

Fig. 7. Direct repeats within the coding sequence of chorion genes which are 13 long. There are 8 bp separating the two repeats. The C to T transition is a silent change, but the C to G transversion causes the replacement of a cysteine with a serine residue. In addition, the second G following the first substitution has been changed to a C in HC-B.12 (sequence data from Rodakis et al. 1984)

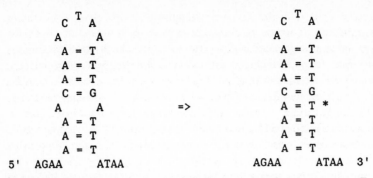

Fig. 8. A palindrome in intron 3 of the *Adh* gene which templates an A to T transversion. The consensus sequence is shown on the *left* and the variant sequence is shown on the *right*. The repeat consist of 8 bp separated by 5. This variant is present in two of eleven sequences (sequence data from Kreitman 1983)

of a serine codon for a cysteine codon. This example is from the chorion gene and the repeats are 13 bp in length separated by only 8 bp. This repeat has had another subsequent mutation, which if included reduces the repeat to a length of nine nucleotides. The expected length of the longest direct repeat anywhere within the sequence has been determined by Karlin et al. (1983). The length of the longest direct repeat (independent of its location) expected within the 1234 bp chorion sequence and given the base composition is 9.27 ± 0.95. Thus, ignoring the feature of their close proximity and ignoring the templating capability of the repeats, direct repeats of 13 bp would still be unusual.

A base substitution which was potentially templated by a nearby palindrome is shown in Fig. 8. The inverted repeat consists of 8 bp separated by 5 bp which templates an A to T transversion in Intron 3 of the *Adh* slow variants found in *D. melanogaster* (Kreitman 1983). The base A is the consensus nucleotide with T as a less common variant.

Selection and the Patterns of Substitutions

The vast majority of the substitutions in the sequences listed in Table 1 are silent or present within flanking or intervening sequences. Very few of the substitutions affect the protein structure. In the intraspecific *Adh* sequences of *D. melanogaster* only 1 of 43 substitutions alter the protein (Kreitman 1983). For the *Adh* sequences of *D. melanogaster, D. simulans* and *D. mauritiana* there are five substitutions affecting the amino acid code, but at least 29 other changes in the region sequenced by Cohn et al. (1984). For the chorion genes 13 of 65 substitutions alter the proteins. For the gamma genes 1 of 37 and for the actin genes 3 of 76. While this does not imply that the substitutions which do not alter the protein are necessarily selectively neutral, it does make an analysis of their selective properties very difficult.

The sequences in Table 1 (which are very homologous) appear to demonstrate that the greatest selective influence is a constraint to maintain the original functional protein. Most of the mutations which alter the protein code must be deleterious to maintain the low frequency of replacement-type substitutions. Many of the substitutions which do occur, appear to be very conservative changes, for example, replacing one hydrophobic residue with a small side group for another. Similarly, the majority of changes induced in *E. coli* tryptophan genes do not appear to significantly alter function (Schneider et al. 1981; Creighton 1974). Unfortunately, this may only reflect our limited ability to detect subtle differences in protein effectiveness.

In part, these genes were chosen for their close homology. The homology is necessary to have some hope of gaining insight into the mechanism of substitutions, but it also places us in a great predicament. An examination of selection is much easier in more divergent species where the sequence differences affect protein structure, but in these cases an examination of the mechanisms of substitution (the initial causes of this divergence) are difficult because the sequences are of limited homology.

With more differences between the sequences one might expect to be able to detect more of the effects of selection. A source of sequence data which is relatively free of bias is difficult to obtain. Sequences which have a "star" phylogeny (Kimura 1983; a radiation from a single common ancestor) have several advantages in this respect, since they do not require phylogenetic reconstruction. For this reason, the sequences which form a star phylogeny, as collected by Gillespie (1986), were examined. There are two sets of sequences used by Gillespie (1986). One set, the "standard" set, are all sequences from different orders of mammals and their common ancestor dates to the time of the mammalian radiation 80—120 million years ago. The "other" set includes homologous genes from a larger collection of taxa (Table 4b, legend).

While Gillespie (1986) examined rates of divergence. I would like to examine the patterns of substitutions in these genes. It is to be hoped that a better understanding of the origins of these substitutions will help to illuminate such studies dealing with other questions.

One of the notable features of these sequences is a high number of parallel mutations. Since each sequence is a member of a star phylogeny and descended from a single common ancestor, each is supposedly an independent example of sequence divergence. Given the expectations discussed previously, the probability of any particular mutation at any particular site should be equal to that of any other mutation at any other site. These sequences demonstrate that this is not true.

The hemoglobin sequences in Gillespie (1986) permit an evaluation of the number of parallel mutations. To measure the number of parallel substitutions, the average number of species that share a particular substitution was calculated. For example, if five species of a star phylogeny have bases A, T, T, C, C at a particular site, then the avergae number of species which share a substitution is 5/3. Three substitutions in total, with the A substitution being present in only one species, the first T (or C) being present in two species and the third substitution, the other T (or C), present in two species. One would normally expect the average number to be close to one, with each substitution being a unique, independent mutational event. This number

Table 4. The mean number of species which share the same substitution compared to the maximum number observed in 100 or 1000 simulations and the corresponding 5% level determined from the simulation

A Mammalian hemoglobin genes

Gene	Codon position	Obs. mean	5% Level	Maximum	Prob. of Obs.	Number of taxa
Alpha hemoglobin	1	1.28	1.09	1.21	< 0.001	4
	2	1.20	1.08	1.19	< 0.001	
	3	1.20	1.11	1.18	< 0.001	
Beta hemoglobin	1	1.19	1.09	1.18	< 0.001	4
	2	1.11	1.09	1.15	0.028	
	3	1.25	1.11	1.18	< 0.001	

B Including homologous genes from other taxa (includes the "other" set of Gillespie, 1986)

Gene	Codon position	Obs. mean	5% Level	Maximum	Prob. of Obs.	Number of taxa
Alpha hemoglobin	1	1.78	1.29	1.31	< 0.01	6
	2	1.82	1.23	1.28	< 0.01	
	3	1.70	1.43	1.49	< 0.01	
Beta hemoglobin	1	1.26	1.23	1.28	0.010	6
	2	1.22	1.22	1.33	0.034	
	3	1.40	1.30	1.38	< 0.001	
Insulin	1	1.26	1.21	1.34	0.009	5
	2	1.22	1.19	1.32	0.025	
	3	1.40	1.30	1.37	< 0.001	

Cytochrome oxidase 1	1	1.33	1.11	1.12	<0.01	4
	2	1.21	1.09	1.15	<0.01	
	3	1.31	1.20	1.21	<0.01	
Cyrochrome oxidase 2	1	1.27	1.14	1.18	<0.01	4
	2	1.18	1.11	1.16	<0.01	
	3	1.33	1.22	1.26	<0.01	
Cytochrome oxidase 3	1	1.34	1.12	1.15	<0.01	4
	2	1.47	1.09	1.15	<0.01	
	3	1.39	1.21	1.22	<0.01	
ATPase 6	1	1.27	1.15	1.16	<0.01	4
	2	1.45	1.10	1.13	<0.01	
	3	1.25	1.21	1.23	<0.01	
Cytochrome b	1	1.30	1.14	1.18	<0.01	4
	2	1.22	1.11	1.19	<0.01	
	3	1.30	1.21	1.22	<0.01	

The species for each of the comparisons in A were human, mouse, goat, rabbit for alpha hemoglobin and human, mouse, bovine, rabbit for beta hemoglobin. For entries in B they were human, mouse , goat, rabbit, chicken, duck for alpha hemoglobin / human, mouse, bovine, rabbit, chicken for beta hemoglobin / human, dog, rat, monkey, guinea pig for insulin / and for each of the remaining mito-chondrial genes the species were human, mouse, bovine and rat

will increase as more species share the same substitution. To make sure that an excess number of shared substitutions was not due to chance, simulations were conducted with random mutations.

The average number of species sharing substitutions in the hemoglobins are shown in Table 4a. In each case, for each position of the codon, there is an excess of parallel substitutions. In fact, only the mutations of the second codon in beta hemoglobin are even close to the levels expected by chance alone. All of the other substitutions have far more parallel substitutions than found in any of 100 or 1000 simulations. If the "other" set of Gillespie's (1986) sequences are included (a comparison which is not as appropriate), the results of Table 4b are found. Again, in every case there is an excess of parallel mutations and in most cases this excess is very large. The parallel substitutions in Table 4a are not a result of common ancestry (a violation of the star phylogeny assumption) since many substitutions are shared by one group of species at one site, but by another group at another site.

Nothing can more clearly demonstrate the constraints that selection must operate under during the evolution of these sequences. The specific nature of these constraints still eludes us, but its strength is impressive. A possible explanation would be that only some codons are permitted at particular sites and hence, if a mutation occurs at all, it must be to this specific set of codons. Alternatively, each of these species may have to adapt to similar selective pressures and respond in the same way, with a host of parallel substitutions.

Interestingly, this same nonrandomness is also present at the third codon positions. For the beta hemoglobins (Table 4a), the third position shows even more parallel mutations than the second codon position. Again, it is not difficult to guess at a possible explanation. Many amino acids are limited to two alternative triplet codons and even those amino acids with four codons have a nonrandom codon usage (Grantham et al. 1980; Maruyama et al. 1986) and permissible substitutions are restricted to follow this context.

While these explanations are plausible and probably form the major component of the true explanation, a closer examination reveals that more factors may be involved. Many of the parallel changes which alter the protein have similar codons elsewhere in the protein and often these have also undergone replacements, but to completely different amino acids. For example, in the alpha hemoglobins, residue #22 has parallel substitutions of C to G in the second codon position causing two independent alanine to glycine substitutions. But elsewhere in the protein alanine is substituted by prolines, glutamic acids, threonines, serines, leucines, aspartic acids, valines, cysteines and asparagines. Why two parallel mutations when other amino acids may be able to carry out the required function? It is still not difficult to impose a selective argument based on interactions at this specific site, but the argument becomes much more complicated.

Similar problems arise with third position substitutions. While codon usage may be invoked to explain some of the parallel substitutions, it can be noted for some of the amino acids that their pattern of codon use has changed. For example, in alpha hemoglobin of rabbit, 88% of nine glycine codes are GGC, but in mice only 36% of eleven glycine codes are GGC. In rabbit, ten valine codons exclusively use GTG (100%) but for the ten valines in mice, GTG is used only 50% of the time. (The probability that these proportions are equal, assuming normality, is 0.008 [Z=2.39] and 0.005 [Z=2.58], respectively.) If these taxa are sufficiently divergent that even their codon usage has been altered, the argument that this must be the cause of the excess of parallel substitutions loses some credibility.

Even given the codon usage, some of the parallel substitutions at third positions appear unusual. For example, the glycine codon usage for the cytochrome oxidase 1 genes of bovine, human, mouse and rat are 94 GGA, 28 GGT, 45 GGC, 18 GGG. But at residue #351 the codons are GGG, GGT, GGT and GGG respectively. These codons imply parallel mutations between the two rarest codons (the probability of this due to random chance would be 0.032). Similarly, in the ATPase 6 gene there are parallel mutations between the rare codons for leucine; CTC, CTT, CTC and CTT respectively, with codon use 90 CTA, 27 CTT, 16 CTC and 10 CTG (a probability of 0.045). Note in both cases that the parallel mutations are not shared by the more closely related sequences of mouse and rat.

In each case it is no doubt possible, given sufficient imagination to determine a selective interpretation for these results. And on the whole, this explanation is probably correct for most of the parallel substitutions, but it may not be required for all of the substitutions. As we have seen earlier, mutation can do many unusual things on its own. There is good evidence which suggests that extreme hot spots may exist for mutation (Benzer 1961; Coulondre et al. 1978) and these mutations may happen repeatedly to the same nucleotide. Similarly, an expected consequence of mutations templated by direct or inverted repeats would be parallel mutations.

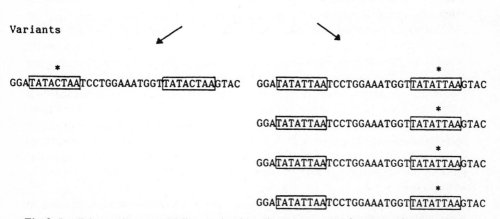

Fig. 9. Parallel mutations potentially templated by direct repeats. The five sequences shown here are from nine secondary response hybridomas. The sequences had several other changes, including others which also occurred in parallel such as shown here for the C to T transition. Both this transition and the complementary T to C transition found in another sequence perfect 8 bp direct repeats separated by 13 bp (After Golding et al. 1987; sequence data from Wysocki et al. 1986)

Unfortunately, the type of data required to determine the plausibility of this mechanism are not abundant. A possible example demonstrating the potential of direct
repeats to template parallel mutations has been found in murine immunoglobulin
sequences (Golding et al. 1987). In this case nine sequences were isolated by Wysocki
et al. (1986) as a response to an immunological challenge. In four of these sequences
the same mutation was found and these mutations perfected the direct repeat shown
in Fig. 9. In addition, the complementary mutation was also found and is included in
Fig. 9. Other parallel mutations can be identified within these genes.

Thus, both selection and nonrandom mutation combine to produce an excess of
parallel substitutions. These types of substitutions imply that correcting for multiple
changes at a single site (as is often required for estimates of divergence) will be a
difficult task.

Discussion

More and more evidence is accumulating that mutations do not occur in the same
fashion as a constant physical process and that the mechanisms of mutagenesis may
be quite complicated (e.g. Drake 1970; Lemontt and Generoso 1982). Many recent
experiments are also demonstrating that there can be an influence of the surrounding
DNA on patterns. This may involve simple effects such as the recognition of a particular sequence by a methylating enzyme. The methylation of cytosine at specified
locations such as "CG" base pairs in vertebrates or methylation of the second adenine
at "GAATTC" sites in *E. coli* can alter the mutational properties of the nucleotides.
The influence of surrounding DNA could also be more complicated and involve local
nucleotide pools or, such presently immeasurable quantitites, as the local three-
dimensional structure. This may prevent the entrance of chemical mutagens or may
change the orientation of the DNA helix. The action of nearby repeats (as discussed
here) is undoubtedly just one mechanism of a multicomponent phenomena.

Evidence is presented here that these nonrandom features of mutation are not
only a feature of the experiments at a laboratory bench, but are also reflected in the
patterns of substitutions that have occurred between individuals and species. There is
definitely a filter between the actual mutations induced and the substitutions observed.
For example, a large preponderance of frameshifts and deletions is seen in spontaneous
mutation (Drake 1970; Schaaper et al. 1986), but is not observed in the substitutions
of coding regions. The reason is the obvious deleterious properties of a frameshift
or deletion in the middle of a protein code. Other kinds of mutations are not as
deleterious and are more clearly reflected in substitutions. For this reason only
some of the mutations caused by direct repeats are likely to also be present as
substitutions.

The number of substitutions that are due to repetitive templates is difficult to
accurately determine. Many potential templates in the sequences discussed here
have not been identified. The limit of 100 bp used for the search is unrealistically
small. In addition, many repeats would not be identified because subsequent fixations
have destroyed the repeats. To be considered here, the repeats must be absolutely

continuous and once a secondary mutation has been fixed in the middle of the repeat, the repeat can no longer be detected. Furthermore, several of the repeats analyzed by Ripley (1982) and Ripley and Glickman (1983) involved repeats that are extensive but not absolutely continuous. Again, these go undetected. The repeats that are found to template mutations in vitro (Kunkel and Alexander 1985) would similarly go undetected since many are too short and/or too distant to be inferred from sequence studies alone. Hence, these results may be underestimates in many ways and it is very difficult to assess what proportion of spontaneous mutations or substitutions may have this underlying cause.

Even given that a small fraction occur via this mechanism, there are other substitutions which most occur as a result of adjacent nucleotides and others as a result of runs of identical nucleotides. The number due to the "classical" mutations, isolated fron context, could by very small. This should also change our expectation that mutations (or substitutions, since these are ultimately dependent) should occur in a consistent fashion or at a constant rate.

Some of these patterns of spontaneous mutations place great genetic constraints on selection. For example, the constant production of small deletions and frameshifts must be continuously counteracted by selection. However, besides imposing some genetic constraints, these mechanisms can also release the organism from constraints. For example, by "classical" mutations (in isolation), if an organism is to adapt to an environmental challenge which requires more than one substitution, it must wait for the first substitution to occur, it must have a still functional product and then it must wait for the second substitution to occur and so on. If mutations are templated by repeats, multiple simulatenous changes can be created and might be the appropriate changes to meet the environmental challenge. This release from constraints, which are normally thought to be universal, presents unique opportunities and a flexibility in adaptive responses that has been previously unsuspected.

The mammalian sequences from a star phylogeny (Table 4a) indicate a large number of parallel substitutions. The majority of these are probably the result of selective constraints, but the mechanisms of mutation that are discussed and those that are illustrated in Fig. 3—9 demonstrate that some may be a simple consequence of mutation.

In addition, the very high frequency of parallel substitutions tends to suggest that these genes may be near saturation for mutational changes (conditional on their present state and function). Some of the patterns of these parallel substitutions are difficult to explain on an individual level. Each of the genes in Table 4a is a hemoglobin, each must carry oxygen and each must adapt to changing conditions. It is difficult to see why rats and man have required the same substitution to adapt to their environment but mice and bovine do not. Surely mice and rats live in more similar niches than do rats and man? In addition, the number of parallel mutations are not abated in degenerate third codon positions. As indicated, the cause need not be selection, nor is mutation likely to be sufficiently strong to be the sole cause of the observed, very high levels. The reason behind so many parallel substitutions in evolution is an interesting question and one well worth the effort. The answer will be a fascinating story.

Summary

Spontaneous mutations can occur by a variety of mechanisms. Some of these mechanisms involve the structure of the surrounding DNA and one such mechanism involving direct repeats and palindromes is examined here. Evolutionary divergence shows an excess number of substitutions which perfect significantly extensive repeats. There is ample availability of less extensive repeats to template a significant fraction of the observed substitutions. This mutational mechanism and others like it demonstrate a greater flexibility than normally expected in DNA sequence changes. While some of the biases present in mutation place constraints on selection (requiring constant correction), other mutational anomalies (such as multiple substitutions) permit a greater range of adaptive responses. The strong selective constraints under which evolution must operate are demonstrated by a clear excess of parallel substitutions. Some of these can be explained by nonrandom mutation.

Acknowledgements. I would like to thank C.H. Langley for this helpful comments on an earlier draft and C. Kelly for the support of her expert typing skills. This work was supported by the Natural Sciences and Engineering Research Council of Canada grant number U0336

References

Albertini AM, Hofer N, Calos MP, Miller JH (1982) On the formation of spontaneous deletions: the importance of short sequence homologies in the generation of large deletions. Cell 29: 319–328

Aoki K, Tateno Y, Takahata N (1981) Estimating evolutionary distance from restriction maps of mitochondrial DNA with arbitrary G+C content. J Mol Evol 18:1–8

Benzer S (1961) On the topography of genetic fine structure. Proc Natl Acad Sci USA 47: 403–415

Bird AP (1980) DNA methylation and the frequency of CpG in animal DNA. Nucl Acids Res 8:1499–1504

Cohn VH, Thompson MA, Moore GP (1984) Nucleotide sequence comparison of the *Adh* gene in three Drosophilids. J Mol Evol 20:31–37

Conkling MA, Koch RE, Drake JW (1980) Determination of mutation rates in bacteriophage T4 by unneighborly base pairs: genetic analysis. J Mol Biol 143:303–315

Couldondre C, Miller JH, Farabaugh PJ, Gilbert W (1978) Molecular basis of base substitution hotspots in *Escherichia coli*. Nature 274:775–780

Creighton TE (1974) The functional significance of the evolutionary divergence between the tryptophan operons of *Escherichia coli* and *Salmonella typhimurium*. J Mol Evol 4:121–137

de Boer JG, Ripley LS (1984) Demonstration of the production of frameshift and base-substitution mutations by quasipalindromic DNA sequences. Proc Natl Acad Sci USA 81:5528–5531

Drake JW (1970) The molecular basis of mutation. Holden-Day, San Francisco

Drake JW, Glickman BW, Ripley LS (1983) Updating the theory of mutation. Am Sci 71:621–630

Fersht AR (1979) Fideliy of replication of phage Phi-X174 DNA by DNA polymerase-III holo-enzyme: spontaneous mutation by misincorporation. Proc Natl Acad Sci USA 76:4946–4950

Foran DR, Johnson PJ, Moore GP (1985) Evolution of two actin genes in the sea urchin *Strongylocentrotus franciscanus*. J Mol Evol 22:108–116

Gillespie JH (1986) Variability of evolutionary rates of DNA. Genetics 113:1077–1091

Gojobori T, Ishii K, Ishii M, Nei M (1982) Estimation of average number of nucleotide substitutions when the rate of substitution varies with nucleotide. J Mol Evol 18:414–423

Golding GB, Glickman BW (1986) Evidence for local DNA influences on patterns of substitutions in the human alpha interferon gene family. Can J Genet Cytol 28:483–496

Golding GB, Gearhart PJ, Glickman BW (1987) Patterns of somatic mutations in immunoglobulin variable genes. Genetics (in press)

Grantham R, Gautier C, Gouvy M (1980) Codon frequencies in 119 individual genes confirm consistent choices of degenerate bases according to genome type. Nucl Acids Res 8:1893–1912

Harris H (1980) The principles of human biochemical genetics, 3rd Edn. Elsevier, North-Holland Biomedical, New York

Josse J, Kaiser AD, Kornberg A (1961) Enzymatic synthesis of deoxyribonucleic acid. VIII. Frequencies of nearest neighbor base sequences in deoxyribonucleic acid. J Biol Chem 236:864–875

Karlin S, Ghandour G, Friedemann O, Tavare S, Korn LJ (1983) New approaches for computer analysis of nucleic acid sequences. Proc Natl Acad Sci USA 80:5660–5664

Kimura M (1980) A simple method for estimating evolutionary rates of base substitutions through comparative studies of nucleotide sequences. J Mol Evol 16:111–120

Kimura M (1981) Estimation of evolutionary distances between homologous nucleotide sequences. Proc Natl Acad Sci USA 78:454–458

Kimura M (1983) The neutral theory of molecular evolution. Cambridge University Press, London

Koch RE (1971) The influence of neighboring base pairs upon base-pair substitution mutation rates. Proc Natl Acad Sci USA 68:773–776

Kornberg A (1980) DNA Replication. WH Freeman, San Francisco

Kreitman M (1983) Nucleotide polymorphism at the alcohol dehydrogenase locus of *Drosophila melanogaster*. Nature 304:412–417

Kunkel TA, Alexander PS (1985) The base substitution fidelity of eucaryotic DNA polymerases. Mispairing frequencies, site preferences, insertion preferences, and base substitution by dislocation. J Biol Chem 261:160–166

Kunkel TA, Schaaper RM, Beckman RA, Loeb LA (1981) On the fidelity of DNA replication: Effect of the next nucleotide on proofreading. J Biol Chem 256:9883–9889

Kunz BA (1982) Genetic effects of deoxyribonucleotide pool imbalances. Environ Mutagen 4:695–725

Kunz BA, Haynes RH (1981) Phenomenology and genetic control of mitotic recombination in yeast. Annu Rev Genet 15:57–89

Lemontt JF, Generoso WM (eds) (1982) Molecular and cellular mechanisms of mutagenesis. Plenum, New York

Maruyama T, Gojobori T, Aota A, Ikemura T (1986) Codon usage tabulated from the GenBank genetic sequence data. Nucl Acids Res 14:151–197

McClelland M, Ivarie R (1982) Asymmetrical distribution of CpG in an 'average' mammalian gene. Nucl Acids Res 10:7865–7877

Müller-Hill B, Kania J (1974) *Lac* repressor can be fused to β-galactosidase. Nature 249:561–563

Ripley LS (1982) Model for the participation of quali-palindromic DNA sequences in frameshift mutation. Proc Natl Acid Sci USA 79:4128–4132

Ripley LS, Glickman BW (1983) Unique self-complementartiy of palindromic sequences provides DNA structural intermediates for mutations. Cold Spring Harbor Symp Quant Biol 47:851–861

Rodakis GC, Lecanidou R, Eickbush TH (1984) Diversity in a chorion multigene family created by tandem duplications and a putative gene conversion event. J Mol Evol 20:265–273

Salser W (1977) Globin mRNA sequences: Analysis of base pairing and evolutionary implications. Cold Spring Harbor Symp Quant Biol 42:985–1002

Schaaper RM, Danforth BN, Glickman BW (1986) Mechanisms of spontaneous mutagenesis: An analysis of the spectrum of spontaneous mutation in the *Escherichia coli lacI* gene. J Mol Biol 189:273–284

Schneider WP, Nichols BP, Yanofsky C (1981) Procedure for production of hybrid genes and proteins and its use in assessing significance of amino acid differences in homologous trypto-phan synthetase alpha polypeptides. Proc Natl Acad Sci USA 78:2169–2173

Slightom JL, Blechl AE, Smithies O (1980) Human fetal Gγ and Aγ globin genes: complete nucleotide sequences suggest that DNA can be exchanged between these duplicated genes. Cell 21:627–638

Smith TF, Ralph WW, Goodman M, Czelusniak J (1985) Codon usage in vertebrate hemoglobins and its implications. Mol Biol Evol 2:390–398

Streisinger G, Okada Y, Emrich J, Newton J, Tsugita A, Terzaghi E, Inouye M (1966) Frameshift mutations and the genetic code. Cold Spring Harbor Symp Quant Biol 31:77–84

Synder RD, Regan JD (1982) DNA repair in normal human and xeroderma pigmentosum group A fibroblasts following treatment with various methanesulfonates and the demonstration of a long-patch (u.v.-like) repair component. Carcinogenesis (Lond) 3:7–14

Takahata N, Kimura M (1981) A model of evolutionary base substitutions and its application with special reference to rapid change of pseudogenes. Genetics 98:641–657

Vogel F, Kopun M (1977) Higher frequencies of transitions among point mutations. J Mol Evol 9:159–180

Watson JD, Crick FHC (1953) The structure of DNA. Cold Spring Harbor Symp Quant Biol 18:123–131

Chapter 9 Genetic Constraints on Plant Adaptive Evolution

B. A. SCHAAL[1] and W. J. LEVERICH[2]

The success of an individual plant is determined by the ability to leave progeny that are themselves successful in reproducing. The determinants of fitness are highly complex in plants and have both genetic and ecological constraints. Fitness is a function of both the genotype of the plant and its interaction with the local micro-habitat. We will consider adaptive evolution here as the process by which populations become better suited to their environment. This includes adaptation to the physical environment, where physiological processes become more closely atuned to such specific habitat features as temperature, moisture, or nutrients. Such evolution also includes adaptation to the biological environment. Among the governing factors here are species-species interaction such as competition, predation, and mutualistic plant-pollinator interactions. Many processes and phenomena constrain or strongly influence adaptive evolution. Factors such as breeding system or genetic covariance among fitness parameters, can limit the range and possibilities of adaptive evolution.

A major nongenetic constraint on adaptive evolution is the phenotypic variation induced by environmental variation. For plants, the environment immediately around the plant is of overwhelming importance in plant adaptation, since individuals cannot easily move to avoid local, unfavorable microhabitats. Plants must adapt to the specific locales in which they find themselves. Because of this, phenotypic plasticity plays an important role in plant adaptation. Levels of phenotypic plasticity are themselves subject to selection (Bradshaw 1965). Any specific phenotypic response is, of course, not heritable and therefore cannot be the basis for further adaptive evolution. Environmental variation, because of its random nature, can confound adaptive evolution. Even a well-adapted genotype cannot persist in a local microsite which does not provide the minimal requirements of the species. On the other hand, a broad range of genotypes can persist if they exhibit phenotypically plastic responses to the environment. Even a marginally adapted genotype can be maintained in a highly favorable microsite. A major issue in plant population biology concerns the relative importance of genotypic variation versus environmentally induced variation in determining plant fitness. Environmental variation in plants clearly constrains adaptive evolution and influences other, genetic constraints.

The genotype of an individual is of major importance in determining fitness, and genetic constraints placed on plant fitness can likewise be complex. How well a genotype allows a plant to conform to and succeed in its specific microenvironment

---

[1] Department of Biology, Washington University, St. Louis, MO 63130, USA
[2] Department of Biology, St. Louis University, St. Louis, MO 63103, USA

Genetic Constraints on Adaptive Evolution
Ed. by V. Loeschcke
© Springer-Verlag Berlin Heidelberg 1987

is considered the essence of adaptation. Such genotype-fitness interactions are mediated, however, by the breeding system of the species. The particular genotypic arrays contained within a plant population are generated by the breeding system of the plant. For example, species which inbreed will have less average heterozygosity than outbred species. Fitness peaks associated with heterozygosity will be more difficult to achieve in these inbreeding plants. Population structure can also influence genotypic determinants of fitness in a manner similar to the breeding system. Plant species which have substructured populations will have different genotypic arrays and crossing relationships than will panmictic populations. In addition to genotypic selection which maximizes adaptation to specific environments, other components of selection can influence and constrain the genotypes available to adaptive evolution. There is increasing evidence for selection at the gamete level, and suggestions of maternal plant selection of gametes or embryos (e.g. Stephenson and Bertin 1983). Such processes can limit adaptive evolution in plant populations by reducing the kinds of genotypes available to natural selection.

What we wish to discuss in the following is some of the experimental evidence from plant studies which indicate genetic constraints on adaptive evolution. Furthermore, we wish to examine the interaction of genetic constraints with environmental variation. First, we will look at the interaction of maternal effect and breeding system in influencing fitness. Next, we examine the role of population structure in determining the fitness of progeny. And finally, the role of sexual selection and its potential influence on adaptive evolution will be briefly considered. Other genetic constraints on adaptive evolution, such as covariance among fitness components during different parts of the life cycle, will be considered elsewhere in this volume.

Breeding System

Plants show a range of breeding systems, from asexual, parthenogenic forms such as *Taraxacum officinale,* to species where outcrossing is enforced by spatial separation of the sexes (Richards 1986). The amount of genetic variation produced by different breeding systems varies. For example, in apomictic forms all progeny of single individuals and all members of a parthenogenic lineage are genetically uniform, although populations of apomictic taxa can be genetically variable because they consist of different asexual lineages (Lyman and Ellstrand 1984). A less extreme restriction of outcrossing occurs in species which are predominantly selfing. In these species there is generally a deficiency of heterozygotes due to inbreeding, but again, populations are often genetically variable (Levin 1978). On the other end of the spectrum of plant breeding systems are obligately outcrossed species. Outcrossing can be enforced by adaptations such as temporal separation of male and female phases or by spatial separation of the sexes as in dioecy (Bawa 1980). Other species are obligately outcrossed due to genetic self-incompatibility systems (Richards 1986). It is common among many plants species, however, to produce seeds by both selfing and outcrossing. Whether or not seeds are the result of inbreeding or outbreeding strongly affects the fitness and, hence, the adaptedness of the offspring.

There is ample evidence from both the population genetics and crop-breeding literature for the superiority of heterozygotes, and for inbreeding depression in predominantly outcrossing species (e.g. Falconer 1960). In such species one might expect to see large differences in the mean fitness of populations, depending on the relative levels of selfing and outcrossing they have experienced. Thus, we expect the breeding system to influence both the distribution of genotypes within populations and the average fitness of a population. Hence, the potential for adaptive evolution will be modified by a population's typical breeding system.

Below we will consider: (1) the relative fitness advantages of progeny produced by inbreeding versus outbreeding; (2) the components of the lifecycle where fitness differences are manifest; and (3) the interaction of genotypically determined fitness with environmental variation.

First, we examine the relative fitness advantages of inbred versus outbred offspring and determine a what stage in the life cycle such differences occur. Fitness consequences of the breeding system were studied in an experimental, greenhouse population of *Lupinus texensis* (Schaal 1984). Seeds were produced by either outcrossing or selfing, and the resultant progeny were monitored throughout their lives in an experimental greenhouse population. In this population there was no effect of the breeding system early in the life cycle (Table 1). Seeds produced by inbreeding as opposed to outcrossing showed no differences in percent seed set, seeds per fruit, seed weight, or time and percent of germination. Plant size after 2 weeks of growth showed no detectable differences between the two progeny groups. However, once these early stages of the life cycle were past, significant differences between inbred and outbred plants were observed. Plant size from 4 weeks after germination was larger for the outcrossed progeny than for the inbred progeny (Table 1). Moreover, survivorship of the outcrossed plants was greater; the total life span is greater for

Table 1. Effect of the breeding system in *Lupinus texensis*

Character	Self	Outcross
Percent seed set	12.5	11.5
Seeds/fruit	4.57	4.56
Seed weight	38.65	39.58
Percent germination	91.55	91.03
Time of germination[a]	1.24	1.17
No. of leaves (2 weeks)	2.62	2.67
No. of leaves (4 weeks)	8.94	10.76[b]
No. of leaves (6 weeks)	20.08	23.44[b]
No. of stems (8 weeks)	5.22	5.56[b]
Mean life span (days)	140	165[b]
Net reproductive rate	88.4	108.7[b]

[a] Mean difference in days
[b] $p < 0.05$

outcrossed plants than for inbred plants (165 days versus 140 days). Total reproduction was also greater for the outbred plants. The average number of flowers per
inbred plant was 88.4, whereas for outcrossed plants it was 108.7 flowers.

This study clearly demonstrates the genetic consequences of breeding system on
the fitness of plants. Plants which are produced by outcrossing have a greater probability of survival and a greater reproductive capacity than do inbred plants. Interestingly, the manifestation of inbreeding depression occurs only after a significant
part of the life cycle is past. Consequently, if natural selection occurs early in the
life cycle, the breeding system in *Lupinus texensis* will not affect adaptive evolution,
since fitness differences are not associated with the breeding system. However, if
natural selection occurs later in the life cycle, then fitness differences among progeny
are associated with genotypes generated by the breeding system.

Now we consider the influence of environmental variation on the genetic constraints placed on adaptive evolution by the breeding system. A significant source
of environmentally induced variation in plants is maternal influences on offspring
fitness. The ability of the maternal plant to nurture developing embryos influences
the fitness of those offspring far beyond the time when they are directly dependent
on the maternal plant. Maternal effects in plants have a very large environmental
component. The more favorable the environment of the maternal plant, the better
able the plant is to provision developing embryos, and the more fit are the resulting
progeny. Maternal effects in plants are often expressed as differences in seed size.
A maternal plant which is in a microhabitat which has ample resources will produce
large seeds, whereas a plant in a poor environment will produce small seeds. The
strong environmental determination of seed size is reflected in the low heritability
of seed size in *Lupinus texensis.* Heritability averages 0.32 in a study of 15 different
populations (Schaal 1985). The resultant fitness consequences of environmentally
induced differences in seed size were examined in the above study of *Lupinus texensis*
which documented the effects of the breeding system (Schaal 1984).

The relationship between the size of seed and various components of fitness
throughout the life cycle is shown in Table 2. In *Lupinus,* seed size shows correlation
with neither seed set nor with seeds per fruit. In this species there is no resource
limited trade-off between seed size and number. Plants in a favorable environment
can produce many, large seeds, whereas plants in poor environments produce few,
small seeds (Schaal 1980). Once the seeds are formed and released from the maternal
plant, seed size has a major influence on fitness. Both the timing of germination
and the percentage of germination are correlated to seed size; small seeds germinate
later and in lower frequency than large seeds (Table 2). The size of plants after
germination is also correlated to seed size. Up to the fourth week subsequent to
germination, there is a positive relationship between seed size and seedling size.
However, by the sixth week after germination there is no longer a significant correlation of plant size to seed size. Fitness components for the rest of the life cycle,
such as total life span and total reproduction, show no detectable correlation to
initial seed size. Maternal effects via seed size are of predominant importance early
in the life cycle of the plant. These data suggest that after the initial life cycle stages,
other factors become more important in determining fitness. It is interesting that
at the same point in the life cycle where maternal effects are no longer apparent,

Table 2. Maternal influences on progeny fitness Correlations between offspring seed size and fitness components

Character	r	p
Percent seed set	− 0.25	n.s.
Seeds/fruit	− 0.36	n.s.
Percent germination	0.49	0.05
Time of germination	0.61	0.05
No. of leaves (2 weeks)	0.76	0.05
No. of leaves (4 weeks)	0.59	0.05
No. of leaves (6 weeks)	0.19	n.s.
No. of stems (8 weeks)	0.02	n.s.
Life span	− 0.18	n.s.
Reproduction	0.28	n.s.

the influences of the breeding system are first detected. Only after maternal effects become diluted are differences in fitness associated with progeny genotype manifested. In this case the genetic constraints on fitness placed by breeding systems are further influenced by environmentally induced variation. Under the circumstances described here, genetic constraints placed on adaptive evolution are limited to specific parts of the plant life cycle.

Next, we consider what is perhaps the most imortant aspect of the breeding system, fitness, and environmental variation. Are fitness differences between inbred and outbred progeny maintained in the face of environmental variation in the field? The above lupine studies were conducted in the greenhouse under optimal conditions and low levels of environmental variation. If the variation found in natural habitats is great, genotypic differences may play a relatively small role in determining fitness. Stochastic, environmental factors associated with the particular microsite may be of overriding importance in the determination of fitness. Schemske (1983) has examined the relative success of plants produced via selfing and outcrossing in three species of *Costus*. The fitness of selfed and outcrossed progeny were measured both in the field and in the greenhouse. The relative influences of environmental variation was determined both in the field and greenhouse. Table 3 shows the fitness of outcrossed and selfed progeny of *Costus allenii* in ints natural habitat in Panama. Total seed production is greater for outcrossed fruits than for selfed fruits. After seed production, however, there are no detectable differences in fitness between outcrossed and selfed progeny. Seed germination and seedling survivorship over a 12-month period are not significantly different for the two progeny groups in the field. The major difference in performance in the field was associated with habitat; the environment has an overwhelming influence on fitness. Seed germination and seedling survivorship were an order of magnitude lower in the shade habitat than in the sun habitat (Table 3). In the greenhouse, however, there were detectable differences in fitness between selfed and outcrossed progeny groups. Mean biomass for out-

Table 3. Field measurements of fitness in *Costus allenii*[a]

Character	Outcrossed	Selfed
Seeds/fruit	45.1	30.5[b]
Seed germination		
Sun	42.3%	31.3%
Shade	4.3%	3.3%
Seedling survivorship		
Sun	11.8%	12.8%
Shade	1.0%	0.0%

[a] From Schemske (1983)
[b] $p < 0.05$

crossed progeny was higher than for selfed progeny in both sun (63.99 vs 549.0 g) and shade (570.3 vs 428.5 g) environments. Moreover, both germination percentage and leaf number had significant variation associated with cross type. In *Costus allenii* there is clear evidence for inbreeding depression, but at least in some years, fitness differences associated with breeding system becomes insignificant in the face of environmental variation. Even within a specific habitat, such as sunny sites, there is enough variation to overcome genetic effects. We do not know how important such relatively small differences in genotypic fitness are in the face of such strong environmental variation. Genotypically based differences in fitness must be important at some point, or adaptive evolution would not occur. Does natural selection occur sporadically over time and space, or are selective differences so small that they are not detected by such studies. Many more plant species need to be studied in the manner of *Costus* before this fundamental question on how evolution occurs in plant populations can be answered. Phenotypic plasticity is extremely important in plant populations and allows plants to adapt to their environments. Thus, much of the variation in fitness among plants in a natural environment may not have a genetic basis and have little influence on adaptive evolution.

Population Structure

Plants can often be distinguished from animals by the major role played by phenotypic plasticity in plant adaptation. In addition, plant populations can also be frequently distinguished from animal populations by their genetic structure. Plant populations may show significant genetic subdivision, where there is significant heterogeneity of gene frequencies within populations. Such genetic heterogeneity can occur either as a selective response to environmental discontinuities (e.g. Jain and Bradshaw 1966), or as a consequence of random genetic drift among sub-

subpopulations (Schaal 1975). We will limit our consideration here to the latter type of genetic structure and evaluate its influence on adaptive evolution.

Population subdivision in plants can result from the mating relations among plants which are imposed by pollinators. Pollen vectors tend to move pollen from a plant to a nearest or nearby neighbor. This limited pollinator movement results in restricted gene flow in plants (Schaal 1980). Continued restriction of gene flow within a plant population over many generations will lead to mating among consanguinous plants and will thereby subdivide a large population into genetic neighborhoods. A neighborhood is defined as the area or number of plants which compose a panmictic unit (Levin and Kerster 1974). When neighborhood size is small, gene frequencies can, via genetic drift, come to differ between neighborhoods, and population subdivision results (Wright 1949). Genetic subdivision due to restricted gene flow leads to population structure characterized as isolation by distance (Wright 1943). Genetic similarity between plants is a function of spatial distance; the closer the distance between two plants, the more similar they are genetically (Fig. 1). An isolation by distance structure has been found in a genetically subdivided population of the perennial plant, *Liatris cyclindracea* (Schaal 1974).

Such an isolation by distance substructure may affect the fitness of the progeny produced within a population. Price and Waser (1979) have postulated differential fitness of progeny based on the geographical distance separating parental plants. Since plants which are close to each other may be consanguinous, the potential arises for inbreeding depression. Likewise, plants that are spatially separated over large distances may be genetically dissimilar; outbreeding depression may occur because of the combining in the F_1 progeny of two disparate genomes. A model of optimal outcrossing is shown in Fig. 2.

Studies of several plant species have suggested that optimal outcrossing may be a factor in some plant populations. Optimal outcrossing has most thoroughly been investigated in the species *Delphinium nelsonii* and *Ipomopsis aggregata* (Waser and

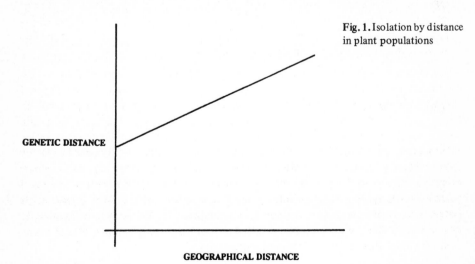

Fig. 1. Isolation by distance in plant populations

GENETIC DISTANCE

GEOGRAPHICAL DISTANCE

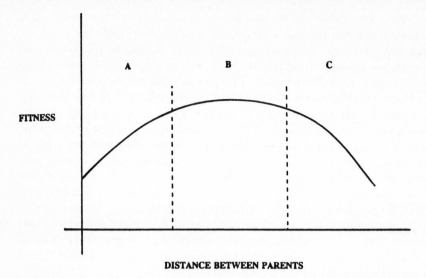

Fig. 2. Optimal outcrossing as a function of spatial separation between parent plants. **A** The region of inbreeding depression; **B** the region of optimal outcrossing; **C** the area of outbreeding depression

Price 1983). Figure 3 shows the relationship of spatial separation between parent plants with several measures of fitness for *Delphinium nelsonii*. The number of seeds per flower conforms to an optimal outcrossing model. Plants which are selfed produce few seeds per flower, presumably due to the homozygosity of deleterious alleles in the selfed progeny. The maximum seed set per flower occurs between 1 and 10 m separation. At greater spatial separation of parent plants the number of seeds per flower declines. Other fitness measures in *Delphinium nelsonii* also vary with distance, but not in strict conformation to optimal outcrossing. Survivorship appears to be negatively correlated to spatial distance between plants, while the proportion of seeds germinating has a general positive correlation. Regardless of whether optimal outcrossing is a general phenomenon, fitness of progeny is clearly a function of distance between parents in *Delphinium*. Thus, the pattern of crosses within a population will influence the fitness of progeny produced. The estimated proportions of crosses within *Delpinium* and *Ipomopsis* populations that are at the optimal distance are shown in Table 4. In both species a large number of crosses will be between plants that are not at potimal separation. Thus, the pattern of mating within these structured populations will produce a number of genotopic arrays with a range of vitness values. Similar results were obtained by Levin (1984) for *Phlox drummondii*. Individual populations did not show strong distance related offspring fitness; however, when all populations were considered together, there was negative relationship spatial separation of parent plants and offspring fitness. In *Phlox* there was no evidence for outbreeding depression. Rather, there is a general increase in seed set with increasing distance. Inbreeding depression in *Phlox* seems to be the major determinant of optimal outcrossing.

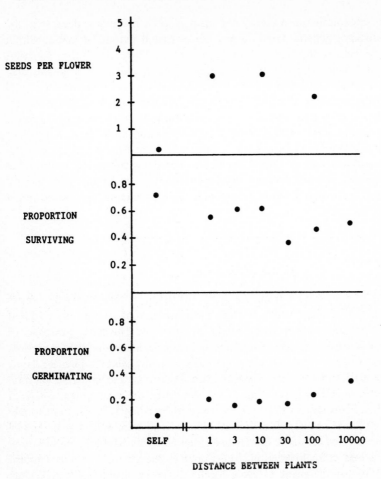

Fig. 3. Fitness of progeny as a function of distance in *Delphinium nelsonii* (After Waser and Price 1983)

Table 4. Pollen transfer and optimal outcrossing interval

	Maximum/offspring fitness distance	Pollinator flights (%)	Flights plus carryover (%)	Dye dispersal (%)
Delphinium nelsonii	1–10 m	7–8	38	6
Ipomopsis aggregata	1–100 m	1–3	9	38

Population substructure, can clearly constrain adaptive evolution. Because of the isolation by distance genetic structure and the restricted pattern of mating within some plant populations, the types of genotypes and the fitness which they represent will be restricted. Homozygosity in substructured plant populations is greater than would be expected under random mating. Thus, the types of genotypes which natural selection can operate on is limited. Furthermore, genotypic arrays within such sub-structured plant populations will also be a function, in part, of the type of pollen vector. Some pollinators, such as butterflies or hummingbirds, have long flight distances, whereas others, such as bees, have very restricted foraging patterns. The proportion of crosses which are within the area of inbreeding depression, outbreeding depression, or at the optimal distance will vary depending on the pollinator species. Thus, the genetic constraints imposed by population substructure have an additional environmental influence.

Conclusions

A major point which emerges from the above discussion is that plant populations have aspects of their biology which strongly influence natural selction and which to some extent distinguish them from animal populations. The basic processes of evolution are, of course, the same for plants and animals. Selection, drift, mutation, migration, and founder events all can potentially influence gene frequencies. However, plant populations have some characteristics which play a greater role in evolution than is seen in animal populations. Variation in the breeding system is a good example of this. Plant species show a wide range of breeding systems, from totally asexual, to sometimes sexual, to obligately bisexual and even multisexual. Animal species show much less diversity; bisexuality is the norm. In addition, various combinations of factors make plant populations unique. Phenotypic plasticity coupled with the inability of the plant to avoid locally unfavorable microsites makes environmental variation of major importance in determining individual fitness. Another example is the combination of isolation by distance population structure with the reliance on insect pollinators which leads to the phenomenon of optimal outcrossing.

The above discussion shows how the plant breeding system, whether it be inbreeding or outbreeding, clearly influences the fitness of progeny and can thereby place limits on adaptive evolution. Plant breeding systems are themselves considered adaptive. The relative benefits of producing genetically variable progeny versus uniform progeny or large numbers of progeny should vary among environments. The observed diversity of plant breeding systems is generally thought to be a result of adaptation to varying environments. Yet, in an immediate context, breeding system can constrain adaptive evolution by limiting or influencing the production of different genotypes with different fitness. Over the long term with such constraints on evolution, one expects the breeding system itself to be modified by natural selection (Bawa 1980).

The influences of genetic population structure may also limit the genotypic array within a population in a similar manner to that by the breeding system. The isolation by distance population structure considered above is a direct consequence of the mutualistic interaction that plants have with their animal pollinators. Since pollinators forage in an energetically efficient manner, it is difficult for a plant species to avoid a population substructure if the species is animal pollinated. Pollinator species could be switched, but those pollinators which do not forage in a restricted pattern are generally species which are a less reliable source of conspecific pollen. Plants have been faced with an evolutionary trade-off between substructured populations with potential constraints on genotypes, on the one hand, and with the securing of efficient and reliable pollination, on the other. Furthermore, the long-term constraints on adaptive evolution of structure are not so straightforward. Sewall Wright (1949), in his shifting balance theory, considered subdivided populations to be the most favorable for genetic change, since new adaptive peaks can be reached within demes via genetic drift. Thus, any proximate genetic constraint on evolution might well be mediated by a long-term advantage. In addition, if inbreeding depression were routinely a constraint in subdivided populations, one expects that eventually natural selection would reduce the genetic load within populations. In fact, there are many plants which routinely inbreed with little fitness liability.

There are several other genetic constraints on natural selection that may be of major importance in plant species. Many of the topics discussed in this volume apply equally well to plants and animals. However, a singular aspect of plants is their alternation of generations. In addition to selection during the diploid phase of the life cycle, selection can also occur during the haploid, gametophytic generation. Such selection can clearly be an additional influence and constraint on natural selection. Moreover, there appears to be increasing evidence for sexual selection in plants (Stephenson and Bertin 1983). Sexual selection may take the form of either gametophytic competition, or mate choice. There is ample evidence for competition among male gametophytes. Competition among pollen grains may be considered a component of natural selection, since a large part of the genome, including recessive deleterious alleles, is expressed and exposed to natural selection. Likewise, there is evidence for female mate choice in plants (Stephenson and Winsor 1986). The progeny of plants which have the opportunity for selective abortion have a greater fitness than do the progeny of plants where selective abortion does not appear to occur. Such sexual selection improves the fitness of progeny early in the life cycle. However, if there are negative genetic covariances between fitness parameters early in the life cycle and fitness at later stages, adaptive evolution will be constrained.

Adaptive evolution in plants is clearly very complex. There is a diversity of genetic constraints such as breeding system, population structure, sexual selection, and negative genetic covariance of fitness among different parts of the life cycle. These all interact. In addition, these processes are all moderated by environmental variation of various types, and in some cases such as maternal effects, several genotypes and environments interact to determine fitness.

Acknowledgments. This work was supported by NSF grant DEB 8141023

References

Bawa KS (1980) Evolution of dioecy in flowering plants. Ann Rev Ecol Syst 11:15–39

Bradshaw AD (1965) Evolutionary significance of phenotypic plasticity in plants. Adv Genet 13:115–155

Falconer DS (1960) Introduction of quantitative genetics. Ronald, New York, 365 pp

Jain SK, Bradshaw AD (1966) Evolutionary divergence among adjacent plant populations. I. Evidence and its theoretical analysis. Heredity 21:407–441

Levin DA (1978) Genetic variation in annual *Phlox:* self-compatible versus self-incompatible species. Evolution 32:245–263

Levin DA (1984) Inbreeding depression and proximity-dependent crossing success in *Phlox drummondii.* Evolution 38:116–127

Levin DA, Kerster HW (1974) Gene flow in seed plants. Evol Biol 7:139–220

Lyman JC, Ellstrand NC (1984) Clonal diversity in *Taraxacum officinale* (Compositae), an apomict. Heredity 53:1–10

Price MV, Waser NM (1979) Pollen dispersal and optimal outcrossing in *Delphinium nelsonii.* Nature 277:294–297

Richards AJ (1986) Plant breeding systems. Allen & Unwin London, 529 pp

Schaal BA (1974) Isolation by distance in *Liatris cylindracea.* Nature 252:703

Schaal BA (1975) Local differentiation and population structure in *Liatris cyclindracea.* Am Nat 109:511–528

Schaal BA (1980) Measurement of gene flow in *Lupinus texensis.* Nature 284:450–451

Schaal BA (1984) Life-history variation, natural selection, and maternal effects in plant populations. In: Dirzo R, Sarukhan J (eds) Perspectives on plant population ecology. Sinauer, Sunderland, Mass, pp 188–206

Schaal BA (1985) Genetic variation in plant populations: from demography to DNA. In: Haeck J, Wodendorp J (eds) Genotypic and phenotypic variation in plant populations. North Holland, Amsterdam, pp 321–342

Schemske DW (1983) Breeding system and habitat effects on fitness components in three neotropical *Costus* (Zingiberaceae). Evolution 37:523–539

Stephenson AG, Bertin RI (1983) Male competition, female choice and sexual selection in plants. In: Real LA (ed) Pollination biology, Academic Press, London, pp 109–149

Stephenson AG, Winsor JA (1986) *Lotus corniculatus* regulates offspring quality through selective fruit abortion. Evolution 40:453–458

Waser NK, Price M (1983) Optimal and actual outcrossing in plants, and the nature of plant-pollinator interactions. In: Jones C, Little R (eds) Handbook of experimental pollination biology. Scientific & Academic Editions, New York, pp 431–459

Wright S (1943) Isolation by distance. Genetics 28:114–138

Wright S (1949) Population structure in evolution. Proc Am Philos Soc 93:471–478

Subject Index

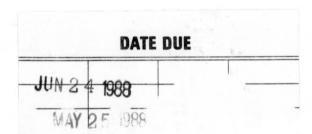